Extreme Nature

Extreme

Nature

Mark Carwardine

with Rosamund Kidman Cox

Collins

An Imprint of HarperCollins*Publishers*

ISBN 978-0-00-724648-9

ISBN 978-0-06-137389-3 (in the United States)

FIRST U.S. EDITION Published 2005

FIRST U.S. PAPERBACK EDITION Published 2008

HarperCollins books may be purchased for educational, business, or sales promotional use. For

information in the United States, please write to: Special Markets Department, HarperCollins

Publishers, 10 East 53rd Street, New York, NY 10022.

The name of the "Smithsonian," "Smithsonian Institution," and the sunburst logo

are registered trademarks of the Smithsonian Institution.

Text © 2005 Mark Carwardine

Color reproduction by Dot Gradation, Essex
Printed and bound in Hong Kong by Printing Express

13 12 11 10 09 08
10 9 8 7 6 5 4 3 2 1

Picture Credits p. 11 © Jeffrey Wood/Royal Botanic Gardens, Kew; p. 12 © Roger Steene/ imagequest3d.com; p. 15 © Thomas Eisner & Daniel Aneshansley/Cornell University; p. 16 © Chris Mattison/FLPA; p. 19 © Ron Toft; pp. 20-1 © Randy Gaugler; p. 22 © James D Watt/ imagequest3d.com; p. 25 © Ken Griffiths/ANTphoto.com; p. 26 © Haroldo Palo Jr/NHPA; p. 29 © Werner A Wuttke; p. 31 © Jeff Lepore/Science Photo Library; p. 32 © Flip Nicklin/ Minden Pictures/FLPA; p. 35 © Matthew Gilligan; p. 36 © Tui De Roy/Minden Pictures/FLPA; p. 39 © Steve Robinson/NHPA; p. 40 © Valerie & Ron Taylor/ardea.com; p. 43 © Mark Moffett/Minden Pictures/FLPA; p. 45 © Mary Plage/ Oxford Scientific Films; p. 46 © Roger Steene/ imagequest3d.com; p. 49 © Mark Carwardine; p. 50 © John Shaw/NHPA; p. 53 © Bill Bachman/ANTphoto.com; p. 54 © ANT Photo Library/NHPA; p. 57 © Mark Carwardine; p. 58 © Marty Snyderman/imagequest3d.com; p. 60 © Neil Lucas/naturepl.com; p. 63 © Don Merton; p. 64 © Peter Batson/imagequest3d.com; pp. 66-7 © Andrea Florence/ardea.com; p. 68 © J M Storey/Carleton University; p. 71 © Jenny Pegg; p. 73 © Raymond Mendez/ Oxford Scientific Films; p. 74 © T Kitchin/V Hurst/NHPA; p. 77 © Peter Batson/imagequest3d.com; p. 79 © David Hosking/FLPA; pp. 80-1 © Robert Suter; p. 82 © Nat Sumanatemeya/imagequest3d.com; p. 85 © Mark Carwardine; p. 89 © Peter Parks/imagequest3d.com; p. 90 © Stephen Dalton/NHPA; pp. 92-3 © Mitsuhiko Imamori/Minden Pictures/FLPA; p. 94 © Flip Nicklin/Minden Pictures/FLPA; p. 97 © Eric Soder/NHPA; p. 98 © Peter Parks/imagequest3d.com; p. 101 © Carlos Villoch/ imagequest3d.com; p. 102 © Christophe Ratier/NHPA; p. 104 © Rob Kay; p. 107 © Alan Williams/NHPA; p. 109 © Chris Mattison/FLPA; p. 110 © Tui De Roy/Minden Pictures/FLPA; p. 113 © Pete Oxford/naturepl.com; p. 115 © Roger Steene/imagequest3d.com; p. 116 © Frans Lanting/Minden Pictures/FLPA; p. 119 © Japan Agency for Marine-Earth Science and Technology (JAMSTEC); p. 120 © Stephen Dalton/NHPA; pp. 122-23 © James Warwick/NHPA; p. 125 © Morley Read/naturepl.com; p. 126 © Mitsuhiko Imamori/Minden Pictures/FLPA; p. 129 © Mark Carwardine; p. 131 © Alan Williams/NHPA; p. 132 © Nigel J Dennis/NHPA; p. 135 © Mark Piazzi; p. 136 © Flip Nicklin/Minden Pictures/FLPA; p. 139 © Ed Reschke/Still Pictures; p. 140 © Rod Planck/NHPA; p. 143 © Pete Oxford/naturepl.com; p. 145 © Michael Fogden; p. 146 © Gouichi Wada/Nature Production/Auscape; p. 149 © Frans Lanting/Minden Pictures/FLPA; p. 151 © Mitsuaki Iwago/Minden Pictures/FLPA; p. 152 © Michael Quinto/Minden Pictures/FLPA; pp. 154-55 © Steve Robinson/NHPA; p. 157 © Michael & Patricia Fogden/Minden Pictures/FLPA; p. 161 © Austin J Stevens/Animals Animal/Oxford Scientific Films; p. 162 © ANT Photo Library/ NHPA; p. 165 © AFLO/naturepl.com; p. 167 © Carl Bento/Australian Museum; p. 168 © Fred Bruemmer/Still Pictures; p. 171 © Jonathan & Angela Scott/NHPA; p. 173 © Ian R MacDonald; p. 174 © S Blair Hedges; p. 177 © Mark Carwardine; p. 178 © Gerard Lacz/FLPA; p. 181 © Frans Lanting/Minden Pictures/FLPA; p. 182 © Erling Svensen; p. 185 © Mark Carwardine; p. 186 © Karl Switak/NHPA; p. 189 © Custom Medical Stock Photo/Science Photo Library; pp. 190-91 © Marc Dantzker; p. 193 © Mitsuhiko Imamori/Minden Pictures/FLPA; p. 194 © James King-Holmes/Science Photo Library; p. 196 © Andrew Parkinson/naturepl.com; p. 199 © Mark Carwardine; p. 201 © ANT Photo Library/NHPA; p. 203 © Tasmanian Herbarium/Tasmanian Museum and Art Gallery; p. 204 © Jim Brandenburg/Minden Pictures/FLPA; p. 207 © Michael Gore/FLPA; p. 208 © Peter Parks/imagequest3d.com; p. 211 © Stephen Krasemann/NHPA; p. 213 © Ann & Steve Toon/NHPA; p. 215 © Flip Nicklin/Minden Pictures/FLPA; p. 216 © Kat Bolstad; p. 219 © Yva Momatiuk/John Eastcott/Minden Pictures/FLPA; p. 220 © Mark Carwardine; p. 222 © Doug Perrine/naturepl.com; p. 225 © Frans Lanting/Minden Pictures/FLPA; p. 226 © Pierre Fidenci; p. 229 © Natural History Museum, London; p. 230 © Pete Oxford/naturepl.com; p. 233 © Winfried Wisniewski/NHPA; p. 234 © Heidi Snell; p. 237 © Merlin Tuttle/Science Photo Library; p. 238 © Science Pictures Ltd/Science Photo Library; p. 241 © Mark Carwardine; p. 242 © Anup Shah/naturepl.com; p. 244 © Peter Herring/imagequest3d.com; p. 247 © K G Preston-Mafham/Premaphotos Wildlife; p. 250 © C Andrew Henley/Auscape; p. 253 © Michael J Tyler, University of Adelaide; p. 255 © Martin Harvey/NHPA; p. 257 © Gabriel Rojo/naturepl.com; p. 259 © Michael & Patricia Fogden/Minden Pictures/FLPA; p. 260 © John Waters/naturepl.com; p. 263 © Frans Lanting/Minden Pictures/FLPA; p. 264 © Konrad Wothe/Minden Pictures/FLPA; p. 267 © Eric Grave/Science Photo Library; p. 268 © Kelvin Aitken/Still Pictures; p. 271 © Alan Chin-Lee; p. 273 © Neil Bromhall/Oxford Scientific Films; p. 274 © Adam White/naturepl.com; pp. 276-77 © Adrian Hepworth/NHPA; p. 279 © Mark Jones/Minden Pictures/FLPA; p. 281 © D Parer & E Parer-Cook/AUSCA/Minden Pictures/FLPA; p. 283 © John Cancalosi/naturepl.com; p. 284 © Yves Lanceau/NHPA; p. 287 © Martin Harvey/ANTphoto.com; p. 288 © Neil Bromhall/ Science Photo Library; p. 291 © Norbert Wu/Minden Pictures/FLPA; p. 293 © Alice & Daniel Harper; p. 294 © Roger Steene/imagequest3d.com; p. 297 © Peter Parks/imagequest3d.com; p. 298 © Frans Lanting/Minden Pictures/FLPA; p. 301 © Andrew Syred/Science Photo Library; p. 302 © Y Kito/imagequest3d.com; p. 305 © Stephen Krasemann/NHPA; pp. 306-07 © Daniel Heuclin/NHPA; p. 308 © Jonathan & Angela Scott/NHPA; pp. 310-11 © Peter Wirtz; p. 312 © Ingo Arndt/naturepl.com; p. 315 © Hideyo Kubota

Contents

Introduction

Extreme Nature is about some of the most intriguing, supernatural, out of the ordinary, and extreme plants and animals on the planet. A fish that can change sex, a frog that gives birth through its mouth, and a flower that smells so bad it makes people faint are in the motley collection of weird and wonderful creatures included in the book.

If proof were ever needed that fact really is stranger than fiction, then look no further than the natural world. Did you know, for example, that a bombardier beetle can blast a chemical spray that's as hot as boiling water? Are you aware that a castor bean produces a toxin 6,000 times more deadly than cyanide? And have you ever wondered about the three-toed sloth, which has just two modes of being: asleep and not quite asleep?

It would have been hard for a science fiction writer to dream up some of the most bizarre creatures and wacky behavior described in this book. Imagine an animal that squirts up to a quarter of its own blood at its predators—that's the Texas horned lizard. There is a frog that can withstand being frozen to $-270°C$ ($-454°F$) and a moth with a 14 in. (35cm) tongue. A fish that can inflate itself to become a spine-covered sphere three times its original size may sound like a character from *The Hitchhiker's Guide to the Galaxy*, but it really does exist, in tropical seas around the world, in the form of the pufferfish.

One of the richest environments for such peculiar and eccentric wildlife is the sea. This is where scientists first discovered the coelacanth, for example—a strange-looking fish that was thought to have gone extinct 65 million years ago, but is now known to be alive and well and living in the western Indian Ocean. The sea is where we first set eyes on a remarkable octopus that lives a life of deception, by disguising itself as anything from a flounder or a jellyfish to a sea snake. And in the cold, dark ocean depths scientists have found shrimp-like creatures thriving 6.8 miles (11 km) below sea level. But the sea is also the most unexplored region of the world and studying its wildlife can be about as difficult and challenging as exploring outer space. Even now, there are probably huge animals lurking beneath the ocean waves as yet unseen by human eyes—and untold numbers of smaller ones. But with the help of space-age research techniques and equipment, such as deep-sea submersibles and remote-access vehicles, we are just beginning to understand the true extent of alien-like life on our own planet.

There are more outlandish creatures to discover on land, too, though many of them are likely to be the natural world's tiniest critters and relatively hard to find. But the fact that we've already unearthed plants that eat animals, insects capable of walking on water, and frogs with baggy skin just makes scientists determined to search for new species of plants and animals and ever-more extravagant forms of behavior. It's hard not to wonder what secrets have yet to be unravelled in the treetops, deep underground, in hidden corners of remote tropical rainforests, or under the glare of a microscope.

Extreme Nature was written with the invaluable help of over 150 such scientists working in all corners of the globe. With their generous assistance, in just a few sentences it's been possible to summarize some of the highlights of many years, sometimes decades, of research. Thanks to them, if you've ever wondered which animal has the best color vision, if a millipede really does have a thousand legs, how fast a falcon can swoop, or which is the world's most dangerous snake, this is the place to look.

Any study of extreme nature is inevitably full of surprises, and when it comes to superlatives, there is always another record-breaker just around the corner. But while there's little doubt that few of the records we've included are absolutes, in one way or another all animals and plants have something exceptional about them that deserves our attention. It's certainly been enormous fun corresponding with so many experts in so many different fields and, with their guidance, making the final selection.

Ultimately, the aim of the book is simply to revel in these other-worldly creatures and their outlandish behavior. We hope you enjoy being wowed by some of their exploits.

Extreme

Abilities

Most devious plant · Strangest boxer · Most explosive defense
Most poisonous animal · Most ingenious toolmaker · Most gruesome
partnership · Best electro-detector · Stickiest skin · Deadliest plant
Biggest blood-sucker · Best sense of smell · Most enthusiastic singer
Most gruesome tongue · Most inquisitive bird · Biggest medicinal drug
user · Most painful stinger · Slipperiest plant · Heaviest drinker · Best
mimic · Most formidable killer · Best architect · Most painful tree · Loudest birdcall
· Deadliest drooler · Most sensitive slasher · Smelliest plant · Most impressive
comeback · Hottest animal · Most shocking animal · Coldest animal · Most
talkative animal · Most bizarre defense · Smelliest animal · Slimiest animal · Best
hearing · Stickiest spitter · Best color vision · Most dangerous snake

Most devious plant

The natural world, as we know it is built on partnerships. But in all societies there are cheats, and plants are no exception. Most green plants would be unable to exist without the help of fungi, which provide them with food-exchange partnerships. In fact, the invasion of the land by plants—algae—was probably only made possible by these types of partnerships. It has even been suggested that early land plants developed roots just so that they could join forces with the fungal roots, or hyphae.

Most plants are real partners, giving the carbohydrates that they manufacture using their chlorophyll. Some—notably orchids—have such a close partnership that they don't even bother to produce food packages to accompany their embryos into the world, instead relying on fungi in the soil to provide the food needed for germination and early growth. This allows an orchid to produce lightweight, microscopic seeds—millions of them.

Some orchids, however, have become cheats: they use fungi that have partnerships with trees, and they never give anything in exchange. Via fungal hyphae, these orchid vampires tap into the trees, siphoning off nutrients. The giveaway is often the fact that they have stopped producing chlorophyll. As a result, they aren't green but a rather sickly pinkish cream, like the ghost orchid, or brown, like the bird's-nest orchid. Some, such as western coralroot, are bloodred or even purple. The drawback is that, without the fungus, the orchid will die. And one day a fungus may just evolve a way to even the score.

NAME **ghost orchid** *Epipogium aphyllum*
LOCATION North and Central Europe eastward to Japan
ABILITY cheating a fungus

Mutual relationships in the sea are common. A famous one is between a hermit crab and a sea anemone, in which the crab gains protection from the anemone's stinging nematocysts, and the sea anemone gets leftover food. Boxer crabs seem to have taken this relationship a stage further. These tiny crabs—their carapace (outer covering) measures only 0.6 in. (1.5cm)—are prey to many animals. Their defense is to wield minute, stinging sea anemones on their specialized claws. A jab with a "glove" can cause pain (even to a human) or death, and it seems to be an effective defense—one boxer crab has even been observed fighting off a blue-ringed octopus. The sea anemones are also used in crab-on-crab boxing. But such fights are ritualistic, and boxing opponents hardly ever touch each other with their sea anemones, grappling instead with their walking legs.

When a growing boxer crab has to molt, it must put down its sea anemones and then grab them again when its new carapace has hardened. If the crab then finds itself with only one sea anemone, it breaks it into two, and the sea anemone obligingly duplicates itself. Surprisingly, the sea anemones don't seem to object to being picked up or brandished in the face of a predator—at least, one hasn't been seen trying to escape. And it's hard to tell what it gets in return other than free travel, but since the boxer crab uses the stinging anemone to stun animals for food, maybe the sea anemone gets enough leftovers to make living on a claw worth it.

Strangest boxer

NAME	**boxer crab**, or **pom pom crab**, *Lybia* species
LOCATION	Indian and Pacific oceans
ABILITY	delivering a punch with anemone gloves

In the world of insects ants can overcome almost anything. But they don't always have it their own way. Bombardier beetles deliver an antiant surprise that is amazingly explosive. An ant, a spider, or any other predator that clamps onto a beetle's leg with hostile intent instantly finds itself blasted with a chemical spray that's | as hot as boiling water.

So how does a small, cold-blooded creature manage to do this? Pure chemistry. In the rear of its abdomen there are two identical glands lying side by side and opening at the abdominal tip. Each gland has an inner chamber containing hydrogen peroxide and hydroquinones and an outer chamber containing catalase and peroxidase. When chemicals in the inner chamber are forced through the outer one, the chemicals react together—effectively creating a bomb.

The resulting vapor, now containing the irritants known as p-benzoquinones, explodes from the end of the abdomen with a bang that's audible to a human and a temperature that's scalding to the would-be predator. Even more amazing, the beetle can rotate its abdomen 270 degrees in any direction, so that it can aim with absolute precision. If 270 degrees isn't enough, it can shoot over its back, hitting a pair of reflectors that will ricochet the spray at the necessary extra angle. Scientists find bombardiers fascinating because they're the only animals known to mix chemicals in order to create an explosion.

Most explosive defense

NAME	**bombardier beetles** Carabidae family
LOCATION	every continent except Antarctica
ABILITY	mixing chemicals to create an explosion

Most poisonous animal

NAME **golden poison dart frog**, or **poison arrow frog**, *Phyllobates terribilis*

LOCATION Pacific rainforests of Colombia, South America

ABILITY producing the deadliest poison of any animal

This tiny frog uses toxic chemicals as a defense in its body, and it is therefore technically poisonous (venomous animals inject toxins via a weapon—a tail, fang, spine, spur, or tooth). The toxin is only effective when the frog is attacked, and since the frog doesn't want to be harmed, its body is a vibrant yellow or orange color to warn predators of extreme danger.

In fact, this most poisonous of frogs is possibly the most poisonous animal in the world. The toxin is in its skin—you can die even just by touching it—and there is enough in the skin of one frog to kill up to 100 people. Although the frog has only been known to science since 1978, inhabiting just one area in Colombia, the Chocó peoples have known about it for generations, using its skin-gland secretions to poison their blowgun darts and kill animals in seconds.

The golden poison dart frog gets most of its batrachotoxin (frog poison) from other animals, probably small beetles, which in turn get it from plant sources. In comparison, captive-bred frogs never become toxic, presumably because they aren't fed toxic insects. The frog is active during the day, since it has few predators except one snake that has become immune to the toxin. Surprisingly, birds have been discovered in New Guinea with the same batrachotoxin in their skin and feathers. The most likely link has been tracked down to a small beetle, similar to the New World beetles, that also contains batrachotoxins.

Most ingenious toolmaker

NAME	**New Caledonian crow** *Corvus moneduloides*
LOCATION	Pacific island of New Caledonia
ABILITY	thinking up ways to get at hard-to-reach food

Several nonhuman animals—from sea otters to woodpecker finches, use tools and sometimes even make them. It's usually assumed that the animal that is the best at this is the chimpanzee. Chimps use rocks to crack open nuts and cut and use twigs or blades of grass to fish in mounds for termites. These are "cultural" skills, practiced only by certain groups of chimps and taught to their youngsters. The skills are certainly sophisticated: an anthropologist once spent several months with a group of chimps trying to learn the art of termite fishing and finally achieved the proficiency, he figured, of a four-year-old chimp novice.

But for pure innate, instant ingenuity, it would be hard to beat the New Caledonian crow. In an experiment in a lab meat was put in a little basket at the bottom of a Perspex cylinder—and a female crow, Betty, was given a straight piece of wire. Holding the wire in her beak, she tried to pull the basket up and failed. Then she took the wire to the side of the box holding the cylinder, poked it behind some tape, and bent the end into a hook. She went back, hooked the basket, and got the meat. When the experiment was repeated, Betty did roughly the same thing, but with the addition of two different toolmaking techniques. In the wild New Caledonia crows make food hooks out of twigs, by snipping off all but one protuberance. But bending a piece of wire . . . how would a crow know?

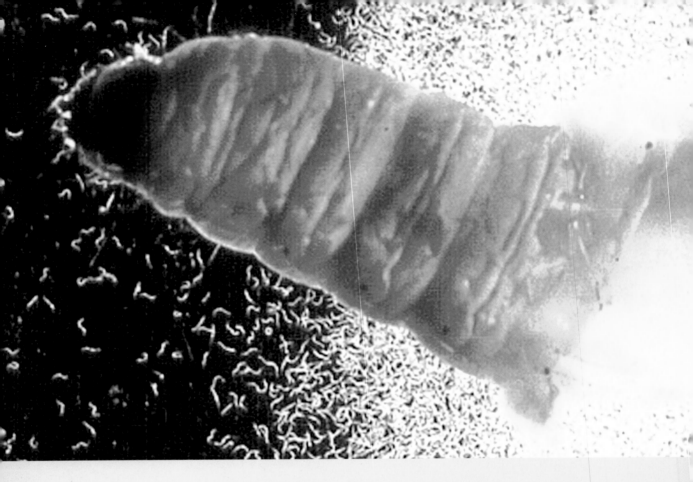

Most gruesome partnership

NAME	**luminescent bacteria** *Photorhabdus luminescens* meets **nematode** *Heterorhabditis bacteriophora*
LOCATION	inside a caterpillar or maggot
ABILITY	combining forces to eat larvae alive

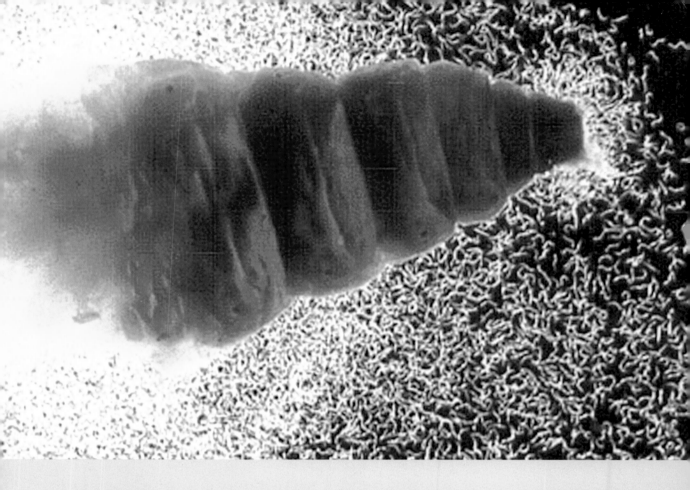

This is a slow and horrible way to die. A wormlike creature, a nematode, goes on the hunt by squirming through the soil in search of an unsuspecting insect grub, or larva. It's not very fussy and likes anything from a weevil to a fly maggot, but it may take months to find a suitable victim. When it does, it penetrates the cuticle of the larva—either through an opening, such as a breathing pore, or by hacking a hole with its special "tooth." Once it is inside, it sets free more than 100 bacteria from inside of its gut, which start producing deadly toxins, digestive enzymes, and antibiotics.

The bacteria are luminescent, and as they multiply, the grub takes on a deathly glow. The nematode then starts feeding on the bacteria and the remains of the corpse of the grub, which is kept free from other competing microorganisms by the antibiotics. Finally, it changes into a hermaphrodite female, laying eggs in the cadaver that will hatch into both male and female nematodes.

Even more of her eggs develop inside of her, and when they hatch, the young nematodes eat their mother. They then mate and produce their own eggs. It's at this point, two weeks later, that the grub finally breaks up, and thousands of young nematodes (each carrying the bacteria in their guts) exit into the soil. The bacteria and the nematode can't exist without each other—a gruesome partnership, but one that humans have joined in, helping the little nematodes spread by deliberately releasing them to hunt down garden pests.

All sharks can, to some extent, detect other creatures by sensing the minuscule amount of electricity that they create, simply by virtue of being alive. In most sharks this sense is principally an adjunct to more dominant senses (usually hearing, smell, and sight) and is especially important in the final split second of an attack. But for hammerhead sharks, it's the main sense, and it could be one reason why their heads are shaped in the strange way that they are.

Sharks have special electrical receptors—hundreds of tiny, dark pores called "ampullae of Lorenzini"—that are filled with a conductive gel that transfers electrical impulses to a nerve ending in each pore. Ordinary sharks have these all over their snout and lower jaw, forming a curious pattern of dark holes resembling a sparse five-o'clock shadow.

But hammerhead sharks also have a mass of them across the underside of their oblong heads, which scan across the sandy seabed like metal detectors, searching for prey that can't be found in any other way —creatures such as stingrays and flatfish that bury themselves, lie still, and usually have no discernable scent.

The hammerhead sharks are able to detect the slight direct currents caused by the interaction between the bodies of their prey and the seawater and the even slighter alternating currents caused by muscle contractions around an animal's heart. The eight species of hammerhead sharks can sense this better than most other sharks, and the biggest of these, the great hammerhead shark, which measures up to 20 ft. (6m) long, may be able to sense it the best of all.

Best electro-detector

NAME	**great hammerhead shark** *Sphyrna mokarran*
LOCATION	tropical and warm temperate seas
SKILL	hunting with electricity

Stickiest skin

Some of the world's strangest creatures are found in Australia—a continent of extremes that gives rise to extreme adaptations. The holy-cross toad lives where many other amphibians cannot: in hot, harsh areas inland, where droughts may last for several years. It uses its strong back legs to burrow down into the soil, where it escapes the heat of the day, and when drought sets in, it survives by digging a chamber a few feet underground in which it estivates (becomes dormant), only emerging when the rains return.

Like its close toad relatives, the holy-cross toad also has unique glands in its skin. If it is disturbed or distressed, these glands release a special secretion that turns into glue. The glue hardens in seconds and has a tensile strength that is five times as strong as other natural glues. This is especially useful if ants attack, as even the biggest ant immediately gets stuck to the toad's skin. And since, like all frogs and toads, it sheds its skin and eats it about once a week, the holy-cross toad has the pleasure of swallowing the ants that try to attack it.

Scientists in Australia are now trying to produce an artificial glue that is as good as the toad's. Holy-cross toad glue will stick plastic, glass, cardboard, and even metal together. More importantly, it can repair splits in cartilage and other body tissues, and therefore it might prove to be a miracle adhesive that will help surgeons repair even the most difficult injuries.

NAME **holy-cross**, **crucifix**, or **Catholic toad**
Notaden bennetti
LOCATION Australia
ABILITY producing an amazing "superglue"

Deadliest plant

The castor-oil plant produces possibly the deadliest plant toxin, 6,000 times deadlier than cyanide, but it has also been known as a wonder plant for thousands of years. The secret and the poison both lie in the seed. More than 50 percent of it comprises a rich oil, but to protect it from being eaten there is ricin—a protein that is toxic to almost all animals (lesser quantities of ricin occur in the leaves). The poison, once it has been ingested, inactivates the key protein-making elements of a cell without which it can't maintain itself and dies.

For humans death is prolonged, ending in convulsions and failure of the liver and other organs. There is no known antidote. The most common cause of poisoning comes from accidentally eating the seeds, but ricin can also be administered in aerosol form, in food or water, or injected—as in the famous case of a dissident Bulgarian journalist. While waiting at a bus stop at Waterloo station in London, England, in 1978, Georgi Markov was murdered by being stabbed with the tip of an umbrella that injected a pellet containing ricin. Widely available and easily produced, ricin could be used for biological warfare.

It is just as easy to extract the seed's valuable oil, however, which has been used for at least 4,000 years as an oil for lamps and soap and also as medicine for a huge array of ailments. Today its uses include high-grade lubricants, textile dyes, printing ink, waxes, polishes, candles, and crayons. In the future its array of protective chemicals may even provide a cure for tumors.

NAME	**castor bean** *Ricinus communis*
LOCATION	worldwide; origin unknown, but probably Ethiopia
ABILITY	producing the deadly poison ricin

No, the biggest bloodsucker isn't a vampire bat. Vampire bats, which are native to the tropical Americas, don't actually suck blood—they lap it up. They find large mammals—most noticeably cows, pigs, or horses—make a cut in their skin and then drink the blood. Not being very big (their average body length is 2.5-3.5 in. (6.5-9cm), a single bat only actually consumes a few tablespoons of blood each night, but because of the anticoagulant in its saliva, the prey keeps bleeding for a while after the bat has flown away.

The world's largest leech, measuring up to 18in. (46cm) does actually suck blood, however, and a very hungry one can take in four times its own body weight before it becomes satiated. Since a large Amazon leech weighs around 1.8oz. (50g)—the record is 2.8 oz. (80g)—that's much more than a few teaspoons of blood. Like the vampire bat, the Amazon leech feeds on large mammals, which it attacks when they enter the water, and it also uses an anticoagulant to keep the animal's blood flowing. But the leech injects an anaesthetic, too, so that the temporary host is unaware of what's happening to it.

All leeches are segmented worms —their closest relatives are earthworms—and all, regardless of their size, have precisely 32 segments. A few segments at each end of the Amazon leech are modified into suckers for attaching to prey, and every segment has its own independent nerve center— meaning a leech has 32 brains.

Biggest bloodsucker

NAME	**Amazon leech** *Haementeria ghilianii*
LOCATION	Amazon basin
ABILITY	drinking up to four times its weight in blood

Best sense of smell

Many animals rely on their sense of smell to find food or a mate and even to find their way around. Some live in places where other senses are of little use—eyes don't help much if you spend most of your life in the dark, and ears would be hopeless in an especially noisy environment—so they rely on smell more than most others.

Some animals, such as sharks, are selective in their smelling abilities and are extrasensitive to significant smells that are relevant to activities such as feeding or breeding. In fact, smell is so important to sharks that they have been dubbed "swimming noses." Their smell receptors are fine-tuned to picking up small concentrations of fish extract, blood, and other chemicals—but so are the receptors of many other animals. Some catfish have such super-receptors so that they can smell compounds at one part to ten billion parts of water.

The likelihood is, though, that moths are the recordholders, especially the males. They use their antennae to home in on the sex pheromones, or chemical allures, released by females and can even detect if these females are on plants suitable for laying eggs. Some females release deviously small amounts of pheromones, to make sure that only those males with the most highly tuned antennae can follow the trails. The likely recordholder for the best *known* sense of smell is the polyphemus moth: just one pheromone molecule landing on a male's antennae will trigger a response in his brain.

NAME **polyphemus moth** *Antheraea polyphemus*
LOCATION North America
ABILITY a male can detect a female based on the evidence of just one molecule

Drop a hydrophone into the water in an area where humpback whales are breeding, and you may hear a baffling medley of moans, groans, roars, snores, squeaks, and whistles. These are the hauntingly beautiful sounds made by male humpback whales, which are famous for singing the longest and most complex of all animal songs. Since most singing takes place at the breeding grounds, it is probably used to woo females and to warn away rival males—but the songs may also have more subtle meanings and nuances that we do not yet understand.

A song can last for as long as half an hour, and as soon as the whale has finished, it often goes back to the beginning and sings it all over again. Each song consists of several main components, or phrases, that are always sung in the same order and repeated a number of times but are forever being refined and improved. All the humpback whales in one area sing roughly the same song, incorporating each other's improvisations as they go along. This means that the song that is heard one day is different from the one heard several months later. and, in this way, the entire composition changes over a period of several years.

Meanwhile, humpback whales in other oceans sing very different compositions. They probably all croon about the same trials and tribulations in life, but the differences are so distinctive that experts can tell where a whale was recorded simply by listening to the intricacies of its own special song.

Most enthusiastic singer

NAME	**humpback whale** *Megaptera novaeangliae*
LOCATION	oceans worldwide
ABILITY	singing the longest and most complex songs in the animal kingdom

Most gruesome tongue

This is probably the world's most specialized and gruesome isopod—one of a group of crustaceans including woodlice, marine gribbles, and slaters. Most isopods lead perfectly normal lives as herbivores, scavengers, or carnivores, but some are parasites. *Cymothoa exigua* has a tendency to select the mouth of the spotted rose snapper fish as its hangout.

Latching onto the fish's tongue with its hooked legs (pereopods), it feeds on mucus, blood, and tissue, gradually eating away the tongue. Gripping onto the tongue stub, the isopod then effectively becomes the fish's tongue, growing as its host grows and feeding on particles of meat that float free as the fish eats. The biggest individual isopod recorded was 1.5 in. (39mm), but presumably it can grow to be as big as the fish needs its tongue to be.

Perhaps the practice is not as gruesome as it looks, as the rose snapper can continue to feed, but no one knows whether a time comes when the *Cymothoa exigua* decides to let go and get a taste of blood in someone else's mouth. Strangely, the relationship between the fish and its parasite has been observed only in the Gulf of California, or the Sea of Cortez, although the fish is found in the eastern Pacific, from Mexico to Peru. It is the only known example of a parasite replacing not just a host's organ but also its function (to hold prey)— a hard act to swallow.

NAME **isopod** *Cymothoa exigua*
SIZE up to 0.5 in. (4cm) long
ABILITY eating and then mimicking the tongue of the spotted rose snapper fish

Most inquisitive bird

NAME   **kea** *Nestor notabilis*
LOCATION   New Zealand
ABILITY   curiosity

Parrots are highly inquisitive, but even among parrots, keas are exceptional. They're native to New Zealand's South Island—a cold, snowy, unparrotlike place where keas have to use all their wits to find a meal. While parrots elsewhere fly from one conspicuous fruit to another, keas search under rocks and bark and in bushes, cones, and shells for food such as roots, shoots, berries, or insect larvae. This—plus a mountainous habitat virtually free of predators—has, over 2.5 million years of evolution, made them insatiably curious. And they're especially drawn to things that they've never seen before. So when humans arrived in New Zealand, the keas were delivered a bonanza of new objects to investigate for food.

Nowadays great sources of fascination are camping grounds and ski resorts. These parrots are large and have powerful beaks, so they can rip right through a canvas tent for the sheer joy of investigation. A particular favorite is the rubber found on cars—mostly windshield wipers. One gang of keas is said to have ripped out the rubber lining around the windshield of a tourists' rental car, causing the glass to fall inward and opening up the interior. When the tourists returned, they found clothes, food, and car parts scattered in the snow, while the keas appeared to be playing a game of soccer with an empty Coke can. The birds then retreated and watched—with great curiosity, it seemed —to see what the tourists would do about it.

Biggest medicinal drug user

Yes, humans are the biggest drug users. But we are not the only ones who use drugs, and we are only just beginning to discover the pharmaceutical knowledge of other animals. The current top-of-the-list medicinal drug user is the chimpanzee. Like us, chimps get stomachaches from time to time after overeating or consuming toxins. They also get parasites and diseases, and stressed animals can usually end up feeling pretty sick.

It's not surprising that an intelligent primate such as a chimp, which learns by trial and error and example, would have started to use medicinal products, since their forest habitat is full of them. In Tanzania chimps suffering from diarrhea have been seen using the leaves of the "bitter leaf" tree that local people know as a medicine for malaria, amoebic dysentery, and intestinal worms. Across Africa chimps have been seen seeking out rough-leafed plants, plucking whole leaves from them, carefully folding the leaves, rolling them around in their mouths, and then swallowing them. Excreted whole, the leaves push out parasites such as intestinal worms.

Many other animals also appear to self-medicate. Capuchin monkeys have been seen rubbing their fur with pungent plants that contain healing and insect-repellant properties. Black lemurs rub insect-killing chemicals from millipedes onto their fur. An elephant has been observed seeking out labor-inducing leaves of a tree just before giving birth. Given our increasing need for new antibiotics and remedies, such examples provide a good reason to keep nature's pharmacy intact by respecting the environment.

NAME	**common chimpanzee** *Pan troglodytes*
LOCATION	forests in East, West, and Central Africa
ABILITY	self-medication

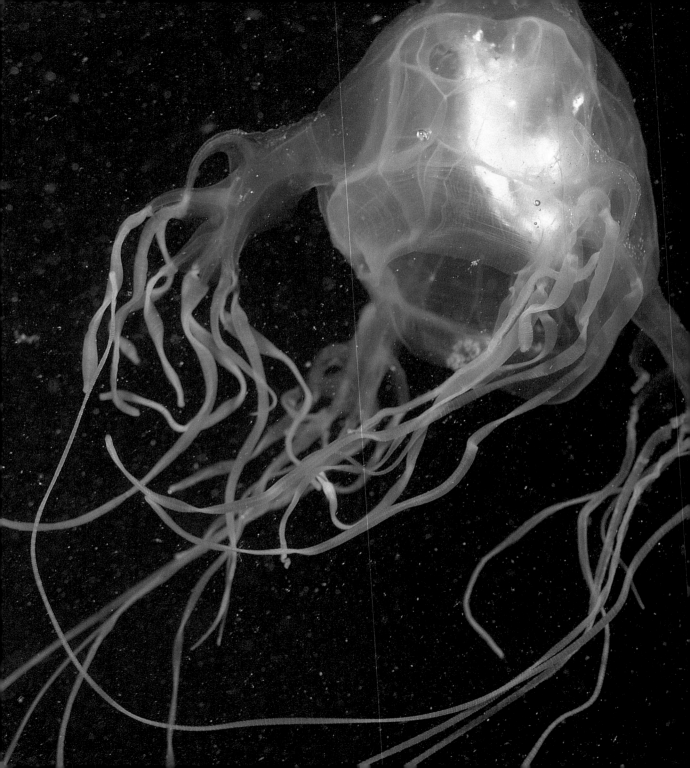

Some say this is the world's most venomous animal, but that depends on what you mean. Is it the venomous creature that you are most likely to encounter, does it kill more people than any other, or are the chemicals the most toxic? Certainly, a single sea wasp contains enough venom to kill at least 60 people, and many do die after being stung.

Although the jellyfish has no desire to kill humans, it is a hunter. An adult sea wasp—as large as a human head, with tentacles up to 15 ft. (4.6m)—has a full array of powerful stinging cells, called nematocysts, and mostly hunts fish. It is very active (unlike many other jellyfish) and jet propels itself through the sea in search of prey. It's also transparent, ensuring that fish (and humans) don't spot its deadly tentacles.

There are four bundles of around ten tentacles, most 6.5 ft. (2m) long and each carrying around three million nematocysts. The toxin contains chemicals that affect heart muscles and nerves and destroy tissue, the purpose being to kill a fish quickly so that it doesn't get away. But if a sea wasp encounters a human, it may also sting in self-defense. The pain is excruciating, and without antivenin, a victim can die from heart failure in just a few minutes. In addition, nematocysts fire not just on command but when they are stimulated physically or chemically. Strangely, they can't penetrate women's tights, and until "stinger suits" became available, lifesavers patrolling beaches would wear tights unashamedly.

Most painful stinger

NAME **sea wasp**, or **box jellyfish** *Chironex fleckeri*
LOCATION near-shore waters of parts of Australia and Southeast Asia
ABILITY causing excruciating pain and possible death

Slipperiest plant

NAME **pitcher plants** *Nepenthes* species
LOCATION Southeast Asia
ABILITY catching prey in slippery, deadly pitchers

There are many different species of pitcher plants, but they are all insect-traps with the slipperiest of sides, providing extra nitrogen (from insect corpses) to help the plants flower and set seed. Among the most sophisticated are the leaves of the vinelike *Nepenthes*. Each of these pitfall traps has an "umbrella" lid and a base that is partially filled with a soup of digestive enzymes. The lure may be color (usually red), smell (nectar or, later, rotting corpses), or tasty hairs. When an insect lands on the rim, it slips into the deadly broth, possibly intoxicated by narcotic nectar.

Slipperiness is achieved in two ways, perhaps depending on which insects are likely to be attracted (walking insects if the *Nepenthes* is on the ground or flying insects if it is up in the tree canopy). The inner walls are usually impossible to climb, since they are covered with slippery, waxy platelets. Others have a surface that attracts a film of water that aquaplanes the insects to their deaths. Some also use trickery. When their pitchers are dry, ants are lured by the nectar, and they don't slip, and so they go and tell more ants about their find. If the surface is wet when they return, they all fall in.

Another of the *Nepenthes* species is in partnership with an ant that has specialized feet, allowing it to get in and out of the pitcher to retrieve corpses. It eats these and drops the remains and its feces into the pitcher, which speeds up the release of nitrogen for its predatory host to ingest.

Heaviest drinker

To say that this or any other hummingbird drinks like a fish is to understate how much it drinks. In proportion to its body weight, it drinks a lot more than a fish. (Just to set that cliché straight: freshwater fish don't drink—they absorb water through their skin. Saltwater fish that drink don't do so to excess.) In the case of the hummingbird, it's the fault of the flowers. Hummingbirds have evolved to drink nectar. The flowers that they visit have evolved to provide that nectar, and the nectar that they provide is typically around 30 percent sugar and the rest is water. To keep their wings moving at a rate quicker than the human eye can see—to hover—hummingbirds need a huge amount of sugar, which means that by drinking nectar they take in up to five times their body weight in water every day.

If any other animal, including a human, tried to drink even one times its body weight, it would be dead long before it could do so. So while hummingbirds were evolving beaks to fit into the flowers with their watered-down nectar, they were also having to evolve nature's heaviest-duty kidneys. Some water just passes through the bird unprocessed, but 80 percent goes to the kidneys to be expelled as very diluted urine. And why the broad-tailed hummingbird in particular? It's simply the most energetic hummingbird, and thus the most supersaturated.

NAME **broad-tailed hummingbird** *Selasphorus platycercus*
LOCATION North America
ABILITY can drink up to five times its body weight in a day

Best mimic

NAME **Indo-Malayan mimic octopus**
LOCATION Indonesian-Malaysian archipelago
ABILITY pretending to be anything but an octopus

If you are a medium-sized predator, the average octopus is one of the most edible animals in the sea. It's substantial and meaty, and without a shell, bones, spines, poisons, or any other unpleasant defense mechanisms. In fact, the best defense that most species of octopus have is to stay hidden as much as possible and do their own hunting at night.

So to find one in full view in shallow water in daylight was a surprise for two Australian underwater photographers, swimming off the Indonesian island of Flores in the early 1990s. Actually, what they saw at first was a flounder. It was only when they looked again that they saw a medium-sized octopus, with all eight of its arms folded and its two eyes staring upward to create the illusion of a fishy body. An octopus has a big brain, excellent eyesight, and the ability to change color and pattern, and this one was using these assets to turn itself into a completely different creature.

Many more of this species have been found since then, and there are now photographs of octopuses that could be said to be morphing into sea snakes (six arms down a hole, and two arms undulating menacingly), hermit crabs, stingrays, crinoids, holothurians, snake eels, brittlestars, ghost crabs, stomatopod (mantis shrimp), blennies, jawfish, jellyfish, lionfish, and sand anemones. And while they mimic, they hunt—producing the spectacle of, say, a flounder suddenly developing an octopod arm, sticking it down a hole, and grabbing whatever's hiding there.

Anything that can attack and kill the largest animal that ever lived, the blue whale (see p.198) has to be the greatest predator ever (apart, of course, from *Homo sapiens*). But blue whales are peaceable creatures with few defenses aside from their size, and so maybe the killer whale qualifies better on the grounds that it can kill the great white shark. At a maximum length of 29.5 ft. (9m), killer whales are the largest members of the dolphin family and among the largest of all predators, but their real edge is that they're pack hunters and work together to subdue large prey.

Several distinct forms are known—residents, transients, and offshores,—each of which differ significantly in appearance, behavior, group size, and diet. The transients are the ones that tend to specialize in larger prey but, perhaps surprisingly, they travel in smaller groups than their fish-eating relatives: fewer than six or seven is pretty typical (fish-eating groups often comprise 15–30 whales). The transients devise different, often ingenious, hunting techniques for different prey. In the Antarctic, for example, they will tip seals and penguins off ice floes and into the mouths of their group mates; and in Patagonia they beach themselves to grab sea lion pups.

When Basque whalers saw killer whales feeding on the carcasses of dead whales, they called them "whale killers," and the name stuck. Many people prefer to use the more politically correct name, orca, but in Latin, *orcus* means "belonging to the kingdom of the dead," and so it's not much better.

Most formidable killer

NAME	**killer whale**, or **orca**, *Orcinus orca*
LOCATION	oceans worldwide
ABILITY	preying on the world's largest animals

Best architect

NAME   **African termite** *Macrotermes bellicosus*
LOCATION   sub-Saharan Africa
ABILITY   creating an air-conditioned, multistory
communal residence

Some 200 species of ants—most famously leaf-cutter ants—farm fungus inside their nests as a source of fast food. So do around 3,500 beetles and 330 species of termites. But out of all these insects, none seems to cultivate a more difficult crop than the African termite, and no crop requires more elaborate technology to maintain it. The staple fungus of African termites grows only on their feces and needs a very precise temperature. Anything above or below 38°F (30.1°C) is too hot or too cold, and every aspect of the construction of the termite mound is part of an effort to keep the temperature exactly right.

The termites always build with mud above a damp pit. They dig at least two long holes down to the water table. They also construct a ten-foot (3m) diameter cellar, around 3.3 ft. (1m) deep, with a thick central pillar that supports the main part of the mound. This houses the queen termite, the nursery, and the fungus farms. On the ceiling of the cellar are thin, circular condensation veins, and around the sides of the mound are ventilation ducts. On top are hollow towers—chimneys—that rise up 20 ft. (6m) above ground level. Every dimension is exactly right for the precise circulation of air and moisture that will keep the fungus at 36°F (30.1°C), no matter what it's like outside. What's more, workers are only a maximum of 0.8 in. (2cm) in size, and so in relative terms, the mound is taller than any human building—the equivalent of 180 stories.

Of course, any tree could fall on you, and plenty of trees are poisonous to eat—but regardless, the trees that cause the most excruciating pain are the ones that you just brush against. These are the stinging trees that are found in several parts of the world, but they are most persistently painful in that land of advanced toxins, Australia. There are six *Dendrocnide* species, two of which—the northern shiny-leafed stinging tree and the southern giant stinging tree— are large, "tree" like trees, and four of which are more like shrubs. Of the six, the worst agony is said to be inflicted by a shrub, the gympie-gympie, but they all hurt a lot.

What at first looks like a layer of fur on all parts of the tree except the roots is really a mass of tiny glass (silicon) fibers containing toxic chemicals. Just a brush against this tree results in the skin being impaled with a scattering of fibers, which act like hypodermic needles and are almost impossible to extract (Australian first-aid kits sometimes include wax hair-removal strips for this purpose). The poison causes burning, itching, swelling, and sometimes blistering that is said to be at its most unbearable soon after contact but can continue causing pain for years. The fibers can penetrate most clothing, and sometimes airborne fibers can be inhaled. Strangely, the stings don't affect all animals. Insects and even some native mammals actually eat the leaves. The animals that suffer tend to be ones that have been introduced to Australia, such as dogs, horses, and humans.

Most painful tree

NAME	**gympie-gympie stinging tree** *Dendrocnide moroides*
LOCATION	Australia
ABILITY	defending itself with toxic chemicals

Loudest birdcall

NAME **kakapo** *Strigops habroptilus*
LOCATION New Zealand
ABILITY producing a booming call that can be heard up to 3 miles (5km) away

Which bird is the loudest depends on who is listening and where they are. The song of a nightingale overcoming the sound of traffic is so loud (90 decibels) that prolonged exposure to it could, theoretically, damage your ears. So could the even louder, more shrill, 115-decibel cry of a male kiwi or the metallic "bonks" of the Central American bellbird, designed to sound through the thick rain forest. But possibly the best long-distance sound to make is a boom.

In Europe the booming record holder is the bittern. But the world record holder is probably the New Zealand kakapo, which is now extinct on the two main islands and, despite great conservation efforts, now numbers fewer than 90 individuals. Every three or four years the normally solitary males gather at traditional kakapo amphitheaters—display grounds with excavated bowls. Here they puff up the air sacs in their chest and belly and start booming at an average of 1,000 times an hour for six to seven hours a night (kakapos are nocturnal, and sound carries best in the colder night air). They do this for three to four months to call in likely mates to witness their dance displays and for mating. But since this giant, flightless parrot is now confined to a handful of offshore islands, few people will ever hear its eerie, "foghorn" boom.

Intriguingly, the booms of Australasian cassowaries are almost as loud but have an added long-distance element: a low-frequency component that is below the range of our hearing (although it can be felt). It's likely that a kakapo boom also contains ultra-low-frequency sound, but its booming is now so rare that it has yet to be thoroughly analyzed.

The Komodo dragon is a renowned giant: the average male is more than 7.5 ft. (2.2m) long, and some measure up to 10 ft. 2 in. (3.1m). The longest lizard of all, however, is its much slimmer relative, the Salvadori monitor from New Guinea, although two thirds of its maximum length of 8 ft. 8 in. (2.7m) is made up by its tail.

But the Komodo dragon is the heaviest lizard of all, with an average weight of 130 lbs. (60kg) and a maximum weight of 176 lbs. (80kg), and it is a fearsome predator. It has large, sharp, serrated teeth for cutting and tearing its prey, but its hidden weapon is its bacteria-laden saliva. Once bitten, a victim may escape, but within a few days it will succumb to infection. The dragon then tracks it down with its acute sense of smell—a sense that also makes it a super-efficient scavenger.

Although it is a giant by today's standards, the Komodo dragon may be a pygmy compared to one of its mainland ancestors (Flores Island supported other "pygmies," including a now-extinct elephant, on which the Komodo dragon is believed to have preyed). In Australia a true giant once existed—the 23 ft. (6.9m), 1,370 lbs. (617kg) monster monitor *Megalania prisca*, which became extinct around 40,000 years ago. The Komodo dragon poses relatively little threat to humans and usually bites only when it is cornered. But *Megalania*, whether or not it was a deadly drooler, would have been a lizard to be very, very afraid of.

Deadliest drooler

NAME **Komodo dragon** *Varanus komodoensis*

LOCATION islands of Komodo, Rinca, Gili Motang, Gili Dasami, and Flores, Indonesia

ABILITY producing dangerous bacteria-laden saliva

A sawfish has external teeth that are set around a sensitive, flat snout—the saw, or rostrum (here shown from the underside). Swung from side to side, the saw can be used as a powerful weapon to slash schools of fish, such as mullet and herring, which it then eats off the seabed. Generally, the sawfish is a slow and peaceable animal, spending its time in shallow, muddy water and raking the mud with its saw for crustaceans and other prey. The saw-teeth get worn down by all this grubbing, but they grow continuously from their bases, so they don't wear out.

Like its close relatives, the rays, it's perfectly camouflaged against the bottom of the sea, and like its more distant relatives, the sharks, it swims in an undulating way. And like both groups, its hard parts are cartilage, not bone, and its teeth are adapted scales. It has another similarity. Using special cells, the "ampullae of Lorenzini," on its saw and head, it can detect electrical fields that are generated by prey.

One problem for females is that they give birth to live saw-babies. But a youngster's saw is covered with a sheath to make birth relatively painless. A much greater problem for all sawfish (possibly seven species in total) is the fact that their coastal waters are being polluted and developed and that they have been overfished to the point where they are all endangered, some critically. A sawfish's saw is also its downfall. Not only has it been sought after as a trophy, but it also fatally entangles the fish themselves in nets.

Most sensitive slasher

NAME	**sawfish** *Pristis* species
LOCATION	shallow, warm coastal waters
ABILITY	using its saw for slashing and sifting

What smells bad to us often doesn't bother other animals. In fact, the scent of the foul-smelling titan arum—the tallest and probably the heaviest of flowering structures—is positively appealing to carrion beetles and bees. Whether its smell is the worst smell to us has still to be tested (there are other contenders for this, including the even bigger giant titan, *A. gigas*). But the titan arum produces a smell that is sufficiently awful to make people faint.

The "flower," or inflorescence, comprises a vase-shaped spathe (petal-like leaf) at least 4 ft. (1.2m) tall, which grows rapidly from a gigantic tuber weighing up to 177 lbs. (80kg). Out of this rises a spadix, a spike with thousands of tiny flowers more than 8 ft. (2.4m) tall, that is so strange it gives the arum its scientific name: "huge deformed penis." The upper part of the spike produces the smell, and to make it travel farther, the spadix generates heat and may steam at night as it pulses its fragrance of ammonia, rotting flesh, and bad eggs for up to eight hours at a time.

This attracts pollinating, carrion-loving insects, but few people have observed the pollination, probably because the plant flowers only every three to ten years and then just for two days. Once the flower dies and hornbills have dispersed its seeds, it's replaced by a gigantic leaf up to 20 ft. (6m) tall, which makes the food so that, one day, the tuber can grow another stinking flower.

Smelliest plant

NAME   **titan arum**, **corpse flower**, or **devil's tongue**, *Amorphophallus titanum*
LOCATION   western Sumatra, Indonesia
ABILITY   pumping out the smell of decomposing flesh

Most impressive comeback

New Zealand's Chatham Islands are believed to have been the last Pacific archipelago to be visited by humans. Yet when people finally did visit, they stayed, and they did a pretty thorough job of doing what humans have always done to islands: stripped them of a lot of their native plants and animals. The Polynesians arrived around 700 years ago, the Europeans came along in the 1790s, and between them they caused the extinction of 26 of the islands' 68 species and subspecies of birds. The main cause was land mammals that were introduced to the island, and among the sufferers from cats and rats in particular was the 6 in. (15cm) endemic black robin.

By 1900 it had disappeared from the two main islands and survived on only Little Mangere—a tiny, windswept stack with sheer cliffs that helped keep predators away but didn't offer the birds much protection from the elements. By 1972 only 18 were left. By 1976, seven.

In the meantime, though, the government had bought nearby Mangere Island and begun to reforest it, and all the birds were moved there. Nevertheless, by 1980 there were just five, with only one breeding pair. But by fostering eggs to other bird species on other islands—which improved the survival chances of the chicks and spurred the breeding female to nest again—conservationists painstakingly sparked the species back to life. Now there are around 250 black robins on Mangere and South East islands, and there are plans to repopulate other islands in the Chathams.

NAME **Chatham Island black robin** *Petroica traversi*

LOCATION Chatham Islands, east of South Island, New Zealand

ACHIEVEMENT increasing its numbers from just five to 250 in 18 years

Hottest animal

NAME   **Pompeii worm** *Alvinella pompejana*
LOCATION   deep-sea hydrothermal vents
ABILITY   withstanding scalding water

The Pompeii worm thrives in large colonies in one of the darkest, deepest, and most hellish places on Earth—close to a geyser of water that is so hot it could melt the worm in a second. It is also subject to a pressure that is great enough to crush a person and doused in a soup of toxic sulfur and heavy metals. Communities of Pompeii worms cling to the sides of "smokers" 1.2–1.9 miles (2–3 km) under the sea. These belching chimneys grow over hydrothermal vents on volcanic mountain ranges, created from the chemicals that precipitate out as 572°F (300°C) vent water meets cold seawater.

To survive on a smoker requires super-worm strategies. For its home, the worm makes a paperlike chemical-and-heat-resistant tube. For a thermal blanket, it "grows" a fleece of filamentous bacteria, feeding it with sugar-rich mucus that is secreted from its back. This blanket may also detoxify the vent fluid in its tube.

Unlike the vent tube worm *Riftia pachyptila*, the Pompeii worm has a gut and "lips" that it extends to "graze" on bacteria that grow on the surface of the colony. But no one knows exactly how it copes with what are the highest temperatures and temperature gradients experienced by any organism aside from bacteria, for although it angles its head (gills, mostly) away from the hottest water, its tail experiences flushes that are hotter than 176°F (80°C). Eager to make use of the Pompeii worm's technology for human endeavors, scientists are now racing against each other to unravel its survival secrets.

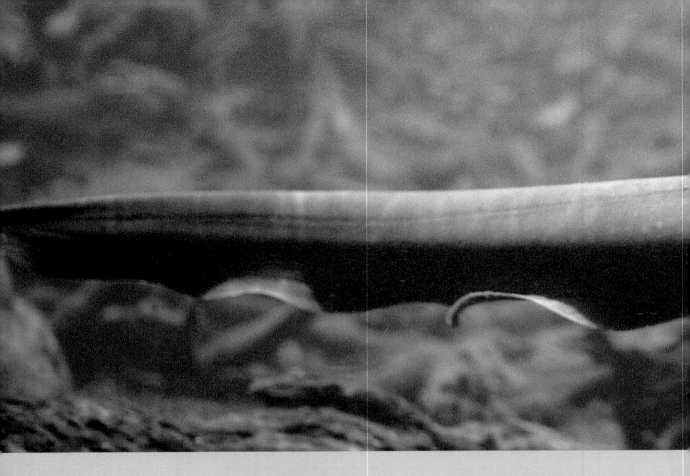

Most shocking animal

NAME	**electric eel** *Electrophorus electricus*
LOCATION	South American rivers
ABILITY	stunning prey with electrical discharges

Think of it as a living battery. The electric eel can grow to more than 6.5 ft. (2m) long, but its organs are right behind its head, leaving 80 percent of its body for the generation of electricity. It's stacked with up to 6,000 specially adapted muscle cells, or electrocytes, aligned like cells in a battery. Each electrocyte emits low-voltage impulses that can add up to 600 volts—enough to render a human unconscious. The positive pole is behind the eel's head, and the negative pole is at the tip of its tail. It tends to remain straight when it swims, using its long ventral fin for propulsion, and so it keeps a uniform electric field around itself.

Electricity affects almost every part of the eel's behavior. As well as stunning or killing with high-voltage pulses, it communicates with other eels electrically and uses electrolocation (a sort of electrical bounce-back system) to detect objects and other creatures in the water. Fish and frogs are its staple prey, and it can detect the minute electric currents that these and other living things produce. The eel can't see well, but this doesn't matter very much, since it is mostly nocturnal and tends to live in murky water.

There are other electrified fish, including the related knifefishes, which generate a weak electric field around themselves that they use to sense objects and fish prey and to communicate. The only other shockers are the torpedo ray and the electric catfish, but neither one is as shocking as the electric eel.

Coldest animal

There's an African midge that is so well adapted to drought conditions that, as a sideline, it can withstand being artificially frozen to $-454\,°F$ ($-270\,°C$). Many other insects can survive freezing, too, but the creatures that can withstand cold for the longest period are probably bacteria in Antarctica.

The most freeze-tolerant "higher" animal is the wood frog, which can literally become a "frog-sicle," enabling it to live farther north than any other amphibian and to hibernate close to snowmelt ponds, presumably to give it a head start and allow it to reproduce quickly before the ponds dry up.

When the temperature drops below freezing, the frog's liver starts converting glycogen into glucose, which acts as an antifreeze. The blood passes the glucose to the vital cells, which are then protected from freezing on the inside, all of the way down to $18\,°F$ ($-8\,°C$). But the rest of the frog's body fluids, up to 65 percent of them, turn into ice and the organs, deprived of blood, actually stop working. Even the eyeballs and the brain freeze. It is effectively the living dead. (The painted turtle *Chrysemys picta* can do this, too, but only briefly.) When a thaw comes, the frog's heart starts beating and pumps blood containing clotting proteins around the body, which stops bleeding from wounds that are caused by the jagged ice crystals. The "frog-sicle" quickly comes back to life and, just as miraculously, so do the frozen parasitic worms in its body.

NAME	**wood frog** *Rana sylvatica*
LOCATION	Canada and Alaska
ABILITY	withstanding being frozen for weeks

Most talkative animal

NAME **African gray parrot** *Psittacus erithacus*

LOCATION central and west Africa

ABILITY communicating with humans

African grays live in huge flocks that sweep through the rainforests foraging for fruit, nuts, seeds, and herbs and constantly communicating with each other. In the wild no one has been able to do more than categorize their calls as, for example, a threat or making contact. But these calls could be much more meaningful and complex if the language skills of pet gray parrots are anything to go by—for African grays can be taught to understand and speak human language and may some day even be able to read words.

The most famous of these parrots (although several others have been making the news recently) is Alex, the protégé of Dr. Irene Pepperberg of Brandeis University in Massachusetts. Alex can identify the colors and shapes of objects and what they're made of. He can, for example, say, "four-corner wood square" if that's what he's been shown. If he wants to be given something or to go somewhere, he only needs to ask. And he can actually make wisecracks and some rudimentary conversation.

One story that Dr. Pepperberg tells about him is how, during a demonstration of his ability to identify alphabetic letters on cards, Alex would say "sssss" for S, "shhh" for SH, "tuh" for T, and so on, and after each correct answer he would ask for a nut. Since that would have slowed things down too much, Dr. Pepperberg each time would say, "Good birdie, but later." Finally Alex looked at her, narrowed his eyes, and said, "Want a nut. Nnn, uh, tuh."

Revered by native peoples for thousands of years, the Texas horned lizard has an array of abilities. It mostly eats ants—and lots of them, since most parts of an ant's body are indigestible. This necessitates a huge stomach. Eating more than 200 ants per day means exposure out in the open for long periods, and having a "heavy" stomach means that a horned lizard finds it difficult to scamper away from predators.

Instead, it relies on an armory of defenses. It has camouflage coloring, with an outline broken up by spines and outgrowths, and it will freeze if a predator approaches. Its horns and spines can pierce the throat of a snake or bird, and it can hiss and blow itself up to look even more fearsome. When it comes to coyotes, foxes, and dogs, a horned lizard's most spectacular defense is to squirt foul-tasting blood from the sinuses behind its eyes. That usually has the desired effect. But it squirts only when it's provoked, since it risks losing up to one quarter of its blood.

Such abilities are, however, no defense against human invasion of its land. Its strange shape and coloring has made it attractive to reptile collectors, and its habit of freezing means that it is prone to being run over. And with human invasion has come an influx of exotic fire ants, that the lizard can't eat. These ants are replacing the native ants on which the lizard depends—a pitiful way for such a determined survivor to go.

Most bizarre defense

NAME	**Texas horned lizard**, or **horny toad** *Phrynosoma cornutum*
LOCATION	southern U.S. and Mexico
ABILITY	squirting blood

Bad smell is all in the nose of the receiver, and we humans have much fewer sensors than most animals do. Nevertheless, we can smell a striped skunk up to 2 miles (3.2km) away, if the wind is in the right direction. It's possible to train our brains to ignore the most disgusting smells—from vomit and feces to rotting flesh—but never skunk. Other animals, including African zorillos, Tasmanian devils, wolverines, and different species of skunks, produce a revolting musk when they are threatened or attacked, but not with the strength or permanence of striped skunk's spray.

The amber oil is made in two muscular glands under the skunk's tail and can be delivered as a spray or precision jet up to a distance of 12 ft. (3.6m). The spray contains compounds that are the cause of the vomit-producing smell, like very, very rotten eggs. It can also cause temporary blindness and, if swallowed, unconsciousness. It is virtually impossible to remove the smell from clothes—and it is usually best to throw them away after a close encounter.

Other mammals are also revolted by it, and so the skunk's only serious predator is the great horned owl, which probably does not have much need for a sense of smell. Skunks prefer not to waste their musk, as the glands take a couple of days to refill, and so they usually raise their black-and-white tail as a warning before spraying. But such warnings go unnoticed on roads, which is why cars are now their worst enemies.

Smelliest animal

NAME	**striped skunk** *Mephitis mephitis*
LOCATION	North America
ABILITY	defending itself with a spray containing the smells that mammals hate most

This is an eel-like animal that is 20–33 in. (0.5–1m) long, without fins, jaws, scales, a backbone, or much of anything in the way of eyes. Though it is not a true fish, it does have gills and excels at a fishlike trait—producing slime. For fish, a thin slime coating is a way of regulating the salt and gas balance between their bodies and the water, repelling parasites, and maintaining speed. But for a hagfish, slime is also a weapon.

Its lifestyle is pretty basic—and even a little squalid: it lives on the seabed, usually at around 4,000 ft. (1,200m), where it eats anything it can overcome or scavenge. When it finds a suitable victim, it slithers into it, usually by way of its mouth, and then it uses its toothed tongue to rip the animal to pieces from the inside out.

That's nothing, however, compared to what it does when it's threatened. Glands all along its sides exude a slime concentrate that reacts with seawater to create a cloud of mucus goo that is hundreds of times larger than the original secretions. It's very tough goo, too—reinforced by thousands of long, thin, strong fibers—and the offending predator or unlucky passerby becomes stuck in it and suffocates. The hagfish itself would suffer a similar fate, but it has a way of extricating itself: it ties itself into a knot and slips the knot down the length of its body, squeezing free in the process.

Slimiest animal

NAME **hagfish** *Myxini* species
LOCATION worldwide
ABILITY drenching predators in mucus by the bucketful

Best
hearing

Hunting and orienting in the dark require an extreme sense. Bats do it by "seeing" with echolocation. They emit high-frequency (ultrasound) pulses from their mouths or noses and analyze the returning echoes to determine the size, shape, texture, location, and movements of the smallest of objects. A bat's nose structure helps it focus the sound, and its complex ear folds catch the returning echoes. Echoes from above hit the folds at different points to those from below. And by moving its ears, a bat can hear sound bouncing back from different angles.

The noise is so intense that, in order to avoid going deaf, most bats "shut off" the sound in their ears as they emit signals. A bat may "shout" at 110 decibels—in the case of the little brown bat—which hunts in open spaces; or it may "whisper" at 60 decibels—in the case of the northern long-eared bat, which catches insects around vegetation. Bats using lower frequencies (longer wavelengths), such as the greater horseshoe bat, tend to either glean insects off vegetation or hunt larger ones; those using higher frequencies (shorter wavelengths) generally catch flying insects at a closer range.

While it is difficult to be certain that the greater horseshoe bat has better hearing than other bats, its echolocation system is one of the few that has been studied in detail, and it is undoubtedly impressive. But many other bats have incredibly precise hearing, too, and it is possible that the real recordholder has yet to be discovered.

NAME	**greater horseshoe bat**
	Rhinolophus ferrumequinum
LOCATION	Europe and Morocco east to Afghanistan and Japan
ABILITY	tracking and catching insects at great speeds in total darkness

Stickiest spitter

Spitting spiders are most closely related to venomous brown recluse spiders. Like brown recluse spiders, they only have six eyes (as opposed to eight) and relatively poor vision. But they make up for it with their snaring skills. Their main sense is touch and, as the spiders walk, their two front legs, which are longer than the other six, tap the ground ahead of them, feeling for something edible.

NAME	**spitting spider** *Scytodes* species
LOCATION	worldwide
ABILITY	snaring and immobilizing prey with sticky spit

Like all spiders, they lay land lines that they periodically tack down with fast-drying silk, which keep them from falling—just like the ropes of a mountain climber. Many spiders can tell if an insect or another spider crosses these lines, and some spiders are believed to use sensitive hairs on their forelegs to detect something without actually touching it—a form of hearing—while others just feel around. But what all spitting spiders do, once their prey has been located, is rear back and spit at it.

With unerring accuracy over a distance that can be more than five times its body length, a spider ejects a gooey mixture of silk and venom. This stuns and immobilizes the prey so that the spider can scurry over, bite it to inject more venom, and eat it. One probable reason for killing at a distance is that spitting spiders are comparatively small and their jaws don't open wide enough to do more than bite a leg or an antenna. The netlike sticky spit also enables them to snare prey that moves faster than they do.

Best color vision

NAME **stomatopods**, or **mantis shrimps**, in the order Stomatopoda
LOCATION nearshore waters worldwide
ABILITY seeing colors you can't even imagine

A stomatopod's main claim to fame is the power of its spring-loaded, armlike dactyls: one boxing blow from a "smasher" stomatopod can punch through the armor of a shellfish. Less famous, but probably even more spectacular, is their range of color vision, especially among the stomatopod species that inhabit the clear shallows around kaleidoscopic coral reefs.

The way that scientists compare color vision in different types of animals is by counting the color photoreceptors in their eyes. Most mammals, for example, have two types. Primates, including humans, are a little better with three, and most birds and reptiles have four. But stomatopods have at least eight and can detect many colors (including thousands of shades of those colors) in the ultraviolet wave band—something we can't see at all. A human can discern around 10,000 colors and shades. A stomatopod can see several times that. This is obviously useful around coral reefs, where many creatures use color as camouflage.

There's also more to their sight than just color. They have polarization vision that is more complex than anything we've been able to create with photography. Their eyes are also on stalks, and they can move independently, scanning 360 degrees. These are compound eyes with thousands of elements, each of which helps see in depth by perceiving an object from three angles. While we need our entire optical setup for one "take" in binocular vision, a stomatopod has much more precise, trinocular vision in a single eye.

Most dangerous snake

This all depends on how you measure danger. Whatever is the best at killing you is the most dangerous. Luckily for humans, no snakes want to eat us, but some can kill when they are defending themselves. The one that kills the most often is the saw-scaled viper. The one with the most toxic venom, however, is the sea snake *Hydrophis belcheri.* Like all sea snakes, its venom has evolved to incapacitate fish and other marine creatures, and it's nonaggressive, lacks the striking fangs of vipers, and bites people only when it is accidentally handled in fishing nets. More dangerous in terms of fatalities is the beaked sea snake, which inhabits coastal waters and so comes into contact with people more often. Many sea snakes are found in Australian waters, and Australia is also the country with the greatest number of venomous snakes. Eleven of the top 12 most venomous snakes are found here—the world record holder being the fierce snake, or inland taipan.

But Australia doesn't hold the record for the most dangerous of all land snakes. Taking into account venom toxicity, venom yield, fang length, temperament, and frequency of bite, the record goes to the saw-scaled viper. It is widespread, small (therefore easily overlooked), and aggressive when it is threatened, and it probably bites and kills more people than any other snake. Its name comes from the fact that, when it is frightened, it rubs its scales together, making a sawing noise—a reminder that most snakes would rather frighten people away than bite them. And, of course, many, many more snakes are killed by people than vice versa.

NAME **saw-scaled viper** *Echis carinatus*
LOCATION West Africa through the Middle East to India
ABILITY injecting a venom that can kill more people than that of any other snake

Extreme

Movement

Strangest hitchhiker · Highest jumper · Largest swarms · Deepest diver
Longest nonstop flier · Most upside-down animal · Most inflatable animal
Heaviest flying animal · Strangest shape-shifter · Fastest flying animal
Greatest sticking power · Largest flying scavenger · Most legs · Speediest sucker
Best long-distance swimmer · Deepest-living animal · Fastest-moving plant
Greatest gathering of birds · Best boring animal · Strongest animal · Best surfer
Earliest-rising songster · Fastest digger · Longest toes · Fastest swimmer
Best water walker · Fastest long-distance runner · Most elastic tongue
Scariest scamperers · Best glider · Biggest sleepover · Fastest sprinter
Longest migration · Biggest hunting army · Sleepiest animal

When you are one of nature's smallest creatures, you have to be ingenious in order to get around. Many tiny arthropods (a group of invertebrates, including insects, crustaceans, spiders, and other similar creatures) choose to hitch a ride to new pastures. The larva of the California spiny lobster starts life at 0.06 in. (3.8cm) and spends months traveling in the open ocean, undergoing 11 molts before it gets to be lobster-sized. To escape from predators, it moves down into deeper waters during the day using its long appendages. But in order to move the hundreds, if not thousands of miles necessary to get to its final, adult feeding place, it hitches a ride on a planktonic jellyfish. Making use of the jellyfish's bell-propulsion sailing mechanism means that the little lobster no longer has to rely on ocean currents.

Other hitchhikers are bolder. The mite *Antennophorus grandis,* for example, chooses to hang onto the underside of the head of a "nurse" yellow meadow ant, which it matches in size. When it's hungry, it strokes its carrier's head, mimicking a demand for a drop of regurgitated sugary food that sister ants make on nurses. So it gets free food as well as a place to hang out. The goose barnacle *Alepas pacifica* goes one step further. It hitches a ride with a female jellyfish but snacks the whole time on the animal's ovaries. The trouble with being a parasitic freeloader, though, is that you are completely dependent on your chosen host. And if you get too greedy, it could be the end for both of you.

Strangest hitchhiker

NAME **larva of the spiny lobster** *Panulirus* species
LOCATION   Pacific Ocean
ABILITY   piggyback riding

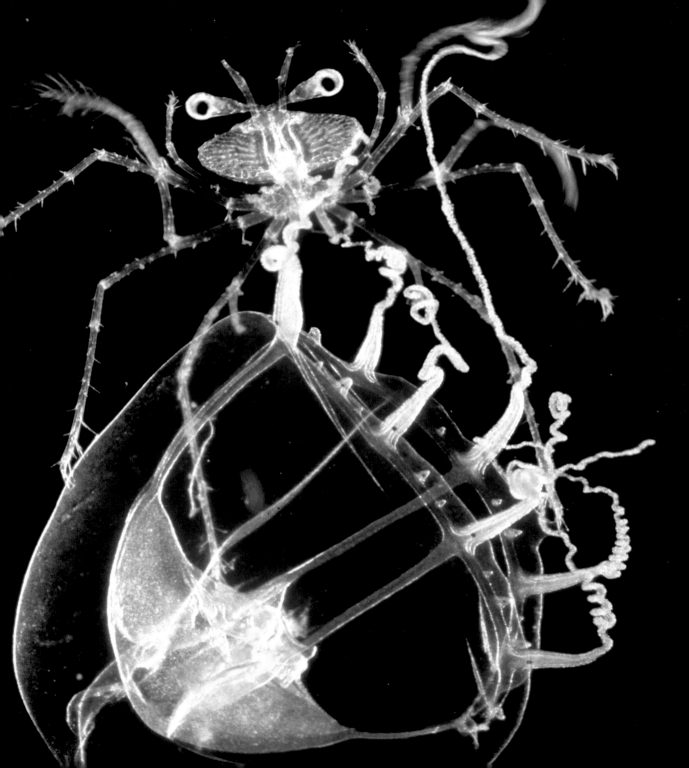

Highest jumper

NAME **spittlebug**, or **leafhopper** species
LOCATION North America and Eurasia
ABILITY jumping no more than 27.5 in. (70cm)

The insects thought of as the most adapted for a life of jumping are fleas—the champion supposedly being the common cat flea, with a jump height of up to 9.5 in. (24cm). Such feats enable them to get to high enough feeding places on moving mammals. But other, less obvious recordbeating hoppers easily out-jump them. Spittlebugs are plant-sucking bugs that are capable of flying or jumping to new plants when they need fresh sap. If danger threatens, though, they have an explosive method of escape. The slightest vibration or touch will make these small bugs leap at such a high speed that a collision with your face would hurt.

The huge "thigh" muscles controlling their longest hind legs (hidden under their eletrya, or wing cases) are the giveaway. Special ridges on these legs (similar to Velcro) hold them in a permanently cocked position, while the "thigh" muscles slowly contract, allowing the legs to suddenly unlock and snap open with such force that the bug is catapulted forward. A cuckoo spit spittlebug accelerates within a millisecond to a takeoff velocity of 13 feet (4 meters) a second, generating a g-force of more than 400 gravities (a human going into orbit in a space rocket will probably experience no more than around five gravities). The force equates to more than 400 times the spittlebug's body weight. By comparison, an ordinary flea is far down the scale at 135 times its body weight. But fleas deserve some sort of record, even if it's just for making high-jumping into a lifestyle.

Largest swarms

NAME	**desert locust** *Schistocerca gregaria*
LOCATION	northern and eastern Africa, Middle East, southwest Asia
SIZE	billions, covering huge areas

A huge swarm of Rocky Mountain locusts, which were the scourge of western pioneers in North America, once covered a minimum area equivalent in size to England, Scotland, Wales, and Ireland combined. As it flew over Nebraska, on August 15–25, 1875, it contained an estimated 25–50 million tons of locusts. Mysteriously, the species became extinct in 1902.

These days desert locusts form the biggest and most widely destructive insect swarms. One, measured by a reconnaissance plane in Kenya in 1954, covered 77 sq. miles (200 sq. km). That was only one of several swarms that were in the same general area, covering a total of around 386 sq. miles (1,000 sq. km), reaching an altitude of 0.9 miles (1.5 km), and comprising an estimated 500 billion insects weighing around 98,400 tons.

One of the oddest things about this animal is that it's basically solitary. Most of the time it's a plain, green grasshopper (a locust is essentially a migrating grasshopper). But when the desert conditions change, so do the insects. At times when it's a little wetter than usual, and more grasshoppers than usual are hatching out of their underground eggs, they bump into each other—they touch. At this point they're still larvae, known as hoppers because they can't fly yet. But they are already beginning to swarm because the touching releases a "gregarization pheromone" (a chemical that makes them gregarious and behave as a group). Masses of them march across the ground in the same direction. This stage lasts about a week, and then they reach maturity and start flying—and the plague begins. There's no specific place where the swarms emerge, but if, say, it's the Arabian Peninsula, the locusts—reproducing as they go— can sweep across Africa, destroying every crop in their path.

Deepest diver

NAME	**sperm whale** *Physeter catodon* is the preferred Latin name, but *Physeter macrocephalus* is also applicable
LOCATION	oceans worldwide
ABILITY	diving deeper than any other mammal

Sperm whales behave more like submarines than air-breathing mammals. They disappear into the cold, dark ocean depths to catch deepwater squid or sharks and other large fish.

In 1991, scientists recorded an incredible, record-breaking dive of 6,560ft (2,000m) near the island of Dominica, in the Caribbean. But there is circumstantial evidence to suggest that sperm whales may be able to dive even deeper. On 25 August 1969, for example, a male sperm whale was killed by whalers 160km (100 miles) south of Durban, South Africa. Inside its stomach were two small sharks, which are known to live only on the seafloor. Since the water in that area exceeds a depth of 10,475ft (3,193m) for a radius of some 30-40 miles (48-64km), it is logical to assume that the sperm whale had been to a similar depth when hunting its prey.

The same whale also made one of the longest recorded dives for any mammal. By the time it surfaced to breathe, having caught the two small sharks, it had been underwater for an estimated 1 hour 52 minutes.

Longest nonstop flier

Depending on how you look at it, several birds could qualify for this record. The arctic tern, for instance (see p.152), makes the longest migration—at least 18,600 miles (30,000km) on a yearly round-trip between the Arctic and the Antarctic (and that's only if it goes in a straight line, which it doesn't). Or the wandering albatross (see p.180), which has been tracked by radio on a flight of 9,320 miles (15,000km) lasting 33 days. There's also the tiny rufous hummingbird, just 3.25 in. (10cm) long, which makes one of the longest trips relative to size—an annual 4,340 miles (7,000km) and back between Alaska and Mexico.

But the important point about swifts is that, except for when they're at their nests, they tend just to keep flying. And the reason is simple: they have to. A swift's legs are so stumpy that it can't perch easily—and so it usually doesn't bother. If it ever found itself on the ground, it wouldn't have the leverage to take off again. So a swift can eat, bathe, drink, mate, and even sleep while it flies, and from August, when it leaves its nest in Europe or Asia to fly south, it doesn't necessarily stop flying until it returns again in April.

More importantly, in its first year of life it doesn't come north to nest, which means that it spends practically two years in the air and flies almost the same distance as to the moon and back. Swifts are very busy birds, too. When feeding their chicks, they can make 40 trips per day to the nest, sometimes flying a total of 620 miles (1,000km).

NAME **Eurasian swift**, or **Northern swift**, or **European swift** *Apus apus*
LOCATION Europe and Asia, Middle East, northwest Africa, sub-Saharan Africa
DISTANCE 310,700 miles (500,000km)

Unlike most of its relatives, the blue sea slug is blind and spends its life floating upside down at the surface of the sea, with a raft of fingerlike projections splayed out and air in its stomach for buoyancy. Although it appears passive, it carries deadly defenses—stinging cells—and has predatory intentions. Its beauty is also a disguise: a metallic-sea-blue, upward-facing underside prevents it from being seen by birds looking down, and a silvery-white, downward-facing upside camouflages it against fish looking up.

Sooner or later, it will find itself drawn into an eddy, whirlpool, windrow, or upwelling aggregation of other animals, where it has a chance of bumping into prey. Its favorite food is both large and armed: the Portuguese man-of-war (a colony of polyps) and its smaller, single-polyp floating cousins, the by-the-wind-sailor and the blue button. The moment the blue sea slug touches one, it starts to rasp and suck its way through the animal's soft tissues. Somehow, it eats the tentacles and recycles their stinging nematocysts without firing them off, incorporating them for defense into its own fingerlike extensions. If it bumps into a sea anemone that is attached to floating seaweed, it will dislodge it by grabbing the animal's holdfast and spinning around.

The only other time it abandons its upside-down position is during mating. It's a hermaphrodite, and when a mating couple exchange sperm, they roll around, caressing each other in an embrace that, by comparison with other animals, is one of the most sensual in the sea.

Most upside-down animal

NAME	**blue sea slug** *Glaucus atlanticus*
LOCATION	mostly subtropical and tropical oceans
ABILITY	drifting through life upside down

Most inflatable animal

NAME **puffer fish** *Diodon holocanthus*
LOCATION tropical coastal seas worldwide
ABILITY transforming itself into a spine-covered "balloon"

This could also be the creature with the greatest number of common names. In English alone, these include spiny puffer, porcupine fish, balloon porcupine fish, brown porcupine fish, blotched porcupine fish, freckled porcupine fish, hedgehog fish, puffer, blowfish, globefish, and swellfish—all of which refer to its defensive quills, its ability to expand, or both. When it is relaxed, the puffer fish looks pretty ordinary. But if it's attacked, it can suddenly blow itself up to become a spine-covered sphere that is three times its original size, like a basketball with hundreds of long, thin nails sticking out of it. It does this by taking many rapid gulps of water.

Its stomach, which in the course of evolution has come to be used for nothing else (food is not digested there but is instead passed right into the intestine), is folded in pleats when the fish isn't puffed up. In fact, there are pleats within pleats within pleats—all of the way down to pleats that can only be seen with a microscope.

When the fish perceives danger, it pumps in water, the stomach unfolds, the skin expands, and the scales—which usually lie backward against the skin—spring up as spines. As well as dispensing with the normal function of its stomach, the puffer fish has lost most of its skeleton, for except its backbone (ribs, in particular, would obviously interfere with expansion). A similar type of oral pumping can also be used offensively: the puffer fish's close relative, the triggerfish, sucks in water and then shoots it out again at sea urchins, flipping them over to reveal their soft undersides.

Heaviest flying animal

NAME   **kori bustard** *Ardeotis kori*

LOCATION   eastern and southern Africa

WEIGHT   possibly more than 49 lbs. (22kg)

Bird weight ranges from the 0.06–0.07 oz. (1.6–1.9g) bee hummingbird to the 332–353 lbs. (150–160kg) ostrich (see p.196), and the main difference between the two extremes is that the hummingbird can fly and the ostrich can't. Even if it had functioning wings, the ostrich would be too heavy to fly.

Aerodynamic limitations, involving relative size and strength of muscles, body mass, lift, and thrust put a weight limit on the possibility of an animal taking to the air, let alone using flapping wings for flight. The largest known bird that ever flew was the 177 lbs. (80kg) extinct giant teratorn (see p.111). However, some modern flying birds approach the limit for flapping flight. Swans, wild turkeys, wandering albatross, and Andean condors weigh in the region of 24–35 lbs. (11–15kg). But the recordholder is the kori bustard—and it's not unusual for a male (females are lighter) to weigh 38–42 lbs. (17–19kg).

After a slow and strenuous-looking takeoff, a male bustard can fly, but it seems to prefer not to and instead spends most of its life walking. Flight is mostly reserved for escaping (although the bustard may first attempt to run away) and for local, nomadic migration. But in the case of the excessively heavy and fat individuals (the verified record is 40 lbs. (22kg), it seems unlikely that they can actually fly, and certainly not for any sustained time. There is, of course, a reason for a male bustard choosing to put bulk above flight: height and visibility on the vast African plains are vital for attracting a mate.

Strangest shape-shifter

NAME **cellular slime mold**
Dictyostelium discoideum

LOCATION soil of North America;
biologists' laboratories worldwide

ABILITY changing from a single cell into a
community slug

This single-celled microbe, affectionately known as Dicty, lives a simple, amoeba-like life in soil and leaf litter, hunting down bacteria. It doesn't bother with mating but instead creates clones of itself when times are good. But if food gets scarce or it gets too dry, something strange happens. Dicty secretes a hormone that summons others of its kind, including its clones, to form a much larger, single entity.

Up to 100,000 of these microbes converge and form a type of slug—effectively, a society of individuals covered in a slime sheath. The animals at the front scan ahead and release waves of a hormone to keep the 0.1 in. (2–3mm) stream of individuals moving in the same direction toward the light. Once it is in the air, the "pseudoplasmodium" sits up. Individuals then choose different roles. The front-end ones squeeze back down through the rest of them, anchor, secrete cellulose to form a rigid stem, and die.

Meanwhile, the others (70 percent of the whole) are lifted up or stream up along the stem. At the tip they form a ball, each one secreting a weather-resistant cellulose coat and transforming itself into a spore. The spores are dispersed by the wind, and if conditions are right, they "hatch" back into single-celled microbes. One question that has puzzled scientists is what prompts individuals to opt for a stalk death and sacrifice themselves for the good of the rest, some of which are unrelated to them. They may be chemically tricked into doing so, but they can resist this manipulation. So it would seem that there is give and take, even in the world of slime.

The fastest bird (and, in fact, the fastest-moving wild animal of any type) is certainly a bird of prey and very probably the peregrine falcon. When "stooping," or plunge diving, to catch a bird in midair, a peregrine that weighs just over 2.2 lbs (1kg) could, theoretically, reach a maximum velocity of 239 mph (385kph) in a 4,113 ft. (1,254m) free fall. Although there is a difference between how fast it could fly and how fast it actually does fly, a peregrine diving with a skydiver has been filmed at stooping speeds more than 200 mph (322kph), which is exceedingly close to the theoretical maximum.

However, what's strange about a peregrine's plunge dive is that, when the bird is within 1.1 miles (1.8km) of prey, it will fly in a curved path. Biologists now think that they know why. A peregrine's vision is such that it sees best when its head is 40 degrees to one side, but to angle its head like this at high speeds would cause too much drag. It is therefore faster for a peregrine to dive on a curve because it can hold its head straight while still keeping its prey in view.

But such flight is not conventional flapping flight. The wandering albatross currently holds the record for the fastest sustained flight, with one individual achieving 35 mph (56kph) sustained over more than 497 miles (800km). Albatross, though, use "dynamic soaring," harnessing the power of the wind to glide rather than continuously flapping.

Fastest flying animal

NAME **peregrine falcon** *Falco peregrinus*
LOCATION every continent except for Antarctica
SPEED can be more than 186 mph (300kph) when they are plunge diving

Greatest sticking power

NAME   **geckos** Gekkonidae family
LOCATION   continent-wide, except for Antarctica
ABILITY   clinging onto any surface

Most people would think that geckos hang from the ceiling, or any other surface, by suction or possibly with the help of claws and adhesive. But their sticking power is much more remarkable.

Each foot is covered with half a million microscopic filaments, called setae, from each of which sprout more than a thousand cauliflower-like fibers, known as spatulae. When these spatulae splay out, they are so close to the surface that a tiny charge is generated between their molecules and those of the surface, which draws them to each other—the positive pole of one attracting the negative pole of the other. This is enhanced by the geckos' mechanism of toe uncurling, which enables them to attach and peel off their feet 15 times per second while running. The molecular force is so strong that one foot on glass, with its millions of spatulae all generating molecular force, could support 88 lbs. (40kg).

This mechanism also includes a self-cleaning component: any dirt that is attached to the setae falls off after just a few steps, the attractive forces between the dirt and the setae are less than those between the surface and the dirt. The geckos' cling-on capacity is inspiring technologists to develop everything from self-cleaning, easily detachable adhesive tape to sure-footed microrobots for use in space. However, jumping spiders almost certainly got there first, being able to support up to 170 times their own weight using the same molecular mechanism.

Largest flying scavenger

NAME   **Andean condor** *Vultur gryphus*

LOCATION   Andes, South America

SIZE   a wingspan of up to 10ft. 5in. (3.2m), 18–33 lbs. (8–15kg)

Condors are scavengers, rather than birds of prey, like vultures. Although they occasionally do catch seabirds, they are not designed for speed or stealth. Instead, they spend a lot of their time gliding around in search of dead animals. And the master glider is the largest of them all—the Andean condor. It's perfectly designed for this type of lifestyle: its feet are used as air breaks and for walking and grasping rather than grabbing and killing; its head and neck are relatively feather-free, enabling it to reach into a carcass without messing up its plumage; its large bill is sharp for slicing through hide; and it has strong masticating muscles, enabling it to tear off chunks of meat.

Nevertheless, there is one drawback. The Andean condor can't run well, and with such huge wings, is too heavy and too cumbersome to get airborne just by flapping. So it must live where there are strong updrafts and thermals to assist with takeoff. Once airborne a condor can soar effortlessly, and its primary feathers act as aerfoils, enabling precise turns as it scans for food below. It doesn't have a good sense of smell, so it follows vultures that can sniff out carcasses, and its huge size means that it is at the top of the pecking order when it arrives.

In prehistoric times there existed a South American glider that was twice the size of a condor: the teratorn, the largest-ever flying bird. But this was no scavenger. Its beak could open wide enough to engulf hare-sized prey whole—making it probably the scariest bird ever to have lived.

Although "milli" means one thousand and "pede" means foot, no millipede really has one thousand feet. Unlike centipedes, which have one pair of legs per segment, millipedes appear to have two. But each millipede segment is, in fact, two segments fused together—a smart way of cutting down on bulk while building up legs. Although it's born with only a few segments, a millipede adds more—and therefore more legs—at every molt. Such a huge number of short, muscular legs gives lifting and pushing power, making millipedes slow but strong, unlike the predatory centipedes, which are flat, low to the ground, and fast moving. Most millipedes are also cylindrical and bulldozer-like and burrow for decaying plant matter (making them valuable forest recyclers), preferring to roll up rather than run away from danger.

Each pair of legs moves slightly out of sync with the pairs in front of and behind it, resulting in a wavelike movement of legs along the body. A female millipede finds the rhythmic tapping of hundreds of legs sweeping along her body very sexy, and so the male caresses her as a prelude to mating. He also has a pair of "sex legs," or gonopods, for transferring his package of sperm to her genital opening. Having many legs is handy for cleaning, and most millipedes are fastidious about cleanliness, spending a lot of their free time scrubbing and polishing, even using one pair of legs as brushes for their antennae.

Most legs

NAME	**millipede**
LOCATION	worldwide
NUMBER	up to 750 legs in 375 pairs

A smallish fish spots an even smaller one swimming enticingly slowly toward an innocuous stone. As it darts in the little fish's direction, it feels a whoosh of suction, everything goes black, and that's the last thing it knows. It's been frogfished.

There are 43 species of frogfish, with different colors (and the ability to change color to match their surroundings), different sizes, and different disguises: some look like sponges, some like encrusted stones, some like clumps of seaweed, some like very unfishlike blobs simply hanging in the water. But what they have in common is the ability to look like inanimate and other animate objects. They also have a dorsal fin that has evolved into a fishing pole, complete with a line and artificial lure—which can "wiggle" like a fish, a worm, or a shrimp—and a mouth that can open cavernously wide, suck like the front end of a jet engine, and shut again, all in one sixth of a second.

Frogfish can also be, in their various and ingenious forms of disguise, monstrously ugly. But they didn't evolve to please the sensibilities of humans. What they did turn out to be were animals with the ability to extend their mouths and engulf their victims faster than any other vertebrate predator. (How fast and exactly what they do wasn't even discovered until the invention of high-speed cinematography.) Frogfish are among the few predators capable of swallowing animals whole that are larger than themselves. And they are also among the planet's greatest masters of disguise.

Speediest sucker

NAME	**frogfish** *Antennarius* species
LOCATION	coral reefs worldwide
ABILITY	instantly changing from a stone into a vacuum sucker

The annual round-trip of gray whales is not only an impressive feat of long-distance swimming, but it is probably also the longest regular migration of any population of mammals. Migration takes place between their summer feeding grounds west and north of Alaska and their winter breeding grounds in a series of shallow lagoons along Mexico's Baja California. The gray whales commute along the entire length of the North American coastline and back again every year.

There's one other whale, though, that muddies this claim a little—the humpback whale (see p.32). Humpbacks are mighty migrators, too, although not quite on the scale of the grays, traveling between cool-water feeding grounds and warm-water breeding grounds in both the Atlantic and the Pacific Oceans. They do, however, hold the record for the longest confirmed migration by any individual mammal. In 1990 a humpback whale seen off the Antarctic Peninsula was spotted five months later off Colombia (the whales are recognizable by the black-and-white markings on the undersides of their tail flukes), an impressive journey of 5,175 miles (8,330km) or a 10,350 miles (16,660km) round-trip.

Since then, several other humpbacks have been identified in both places—which confirms that the first one wasn't merely lost. But it's not necessarily a record-breaking migration on the scale of the entire population of the eastern North Pacific gray whales. However, the gray holds one even prouder distinction: the most impressive recovery by a great whale. In the late 1930s, when grays were officially protected, there were only a few hundred left. Now it is estimated that there are a healthy 26,000.

Best long-distance swimmer

NAME	**gray whale** *Eschrichtius robustus*
LOCATION	North Pacific Ocean
ABILITY	swimming up to 9,900 miles (16,000km) per year

Deepest-living animal

NAME   **amphipod** *Hirondellea gigas*

LOCATION   Mariana Trench, Pacific Ocean

ABILITY   thriving 6.8 miles (11km) below sea level at a pressure of 1,000 atmospheres

In 1995, using the only remote-operated vehicle that could withstand pressures 1,000 times greater than the pressure at sea level, scientists finally probed the deepest place on Earth. They explored the Challenger Deep region of the Mariana Trench in the Pacific Ocean, 36,000 ft.(11,000m) below sea level—deeper than Mount Everest is high—and they found bacteria. In particular, they found a cream-colored, sausage-shaped colony of a species later called *Moritella yayanosii*, whose ability to survive such extreme conditions, using special enzymes and proteins, offers the promise of medical breakthroughs.

But they also found amphipods—lots of them. Amphipods are diverse crustaceans that include sand hoppers, and which live on the seabed from the poles to the equator. But one big hopper, at 1.8 in. (4.5cm), was scampering around in one of the most hostile environments on Earth, in absolute darkness, where there is little food and under a pressure that would kill most other animals.

Hirondellea gigas has only been found in the very deep sea. It has a slow metabolism and a large gut, but it can swim surprisingly fast. It survives by scavenging on falls of "marine snow"—the remains of animals that slowly sink to the depths—and must either be plentiful or have a remarkable sense of smell, because when bait is laid down, it invariably turns up. With technology allowing more deep-sea voyages, the likelihood is that more communities of "extreme" animals will be found living at such depths.

Fastest-moving plant

NAME **Venus flytrap**, or **Venus's-flytrap**
Dionaea muscipula
LOCATION bogs in North and South Carolina, U.S.
ABILITY grabbing insects at a speed of a fraction of a second

Plants may have limitations on their movement—and they don't have muscle power—but there are ways around that. The Venus flytrap uses elasticity to set its two-lobed, snap-jaw trap. The animals it kills provide the nitrogen and other essential minerals that it needs in order to stay healthy and set seed in its nutrient-poor, waterlogged, and acidic habitat.

The upper side of each lobe is stretched more tautly than the underside (its cells elongate under water pressure), and the lobe curves back by tension, like a bow. It lures insects with nectar, which it produces via glands along the rims of the lobes. The trap is released when several of the six or so sensitive trigger hairs sense movement and send chemical-electric signals that cause water transfer between cells, releasing the tension. The result is an instantaneous shutting of the trap.

The Venus flytrap should also be admired for another talent. It can "decide" if something inanimate is in the trap, by counting whether there are two or more stimulations of the hairs—any fewer and the trap won't work. The lobes tighten to form a seal as enzymes are released (with antiseptic to deter bacteria and fungi), and then the prey is digested. A week or so later the trap is ready to kill again.

Only bladderworts rival the flytrap for speed. They catch tiny creatures underwater with bladder traps, closing the seals of their traps in a similarly miniscule fraction of a second.

Greatest gathering of birds

NAME	**red-billed quelea** *Quelea quelea*
LOCATION	eastern and southern Africa
SIZE	30 million or more gathered together

The North American passenger pigeon is widely believed to have been the most common bird that has ever lived on Earth. It lived in huge, densely-packed flocks (some containing more than 2,000 million individuals), which darkened the sky and could take up to three days to pass by overhead. But the species was hunted to extinction in less than 100 years, and the last individual died on September 1, 1914.

Today, it is queleas that gather in the world's largest flocks. They're also, as far as anybody can tell, the world's most

numerous birds. The population is so large that it is effectively uncountable—although it is usually guessed to be in the hundreds of millions. Put it this way: there are so many queleas that every year southern Africans can kill 65–180 million of them without appearing to make a dent in the overall population size. And the reason that so many people are so anxious to kill these rather pretty sparrow-sized finches (apart from putting them in a cooking pot) has to do with those greatest bird gatherings.

Queleas are agricultural pests, often called "feathered locusts." When they swarm, they eat, and among the things they eat are cereal crops if there are any around—millet, sorghum, wheat, and rice—and in a short time they can wipe out an entire harvest. The natural food of queleas is wild grass seed, which, as migrants, they find by following the rains. Once they've located a good supply, all the birds from hundreds of miles around gather there, turning trees into masses of feathers and even breaking branches from their combined weight. They will also feed on insects, especially when they are raising chicks, and this actually helps local farmers by clearing out their other crop pests. This should, of course, make agricultural people undecided about queleas. But while most never appreciate the pest-control role, some are aware that, when swarms of queleas appear, it's time for a delicious feast.

A caecilian looks like a cartoon image of an earthworm. The blunt head resembles the tail, except with jaws and sharp teeth, and the body is divided into glistening segments. Caecilians do have a connection to earthworms—many species eat little else—but they are vertebrates (have backbones) and are in fact related to newts and salamanders. However, they have lost their legs, can hardly see, and have very little to do with water. They range from a worm-sized 2.7 in. (7cm) to an improbable 4.9 ft. (1.5m) and are practically never seen by people because most live a subterranean existence.

Caecilians are remarkably good at boring through the ground. Most elongated tunneling animals, such as worms and snakes, use what's called a normal concertina, or accordian, movement—going from closed to open, as the tail muscles push the head and upper body forward. The caecilian, however, always keeps its body straight.

It also appears to go through the ground like a self-propelled iron rod, hammering the very hard earth aside with its head (not that you could ever watch it underground, but scientists in laboratories put caecilians into clear, dirt-filled plastic tubes). In fact, it is using an internal concertina movement. Its vertebral column is flexing and extending, in the same way as a snake's would, but the skeleton isn't connected to the outer sheath of skin. As the skeleton moves, the outer part of the caecilian stays straight, rippling along and giving the animal traction, similar to the treads on a tank.

Best boring animal

NAME **caecilian** species
LOCATION tropical regions
ABILITY hammering its way through hard ground

Rhinoceros beetles belong to the scarab family, many of which are so incredibly strong that they can do things like roll huge dung balls or bury other animals. But the rhinoceros beetle is supposedly the strongest: one laboratory animal reputedly lifted a weight on its back that was 850 times its own weight—far exceeding the relative strength of an elephant.

Even if the record is exaggerated, there is no doubt about this beetle's strength. Males are famous for their forked "horns:" an enormous one arching over the beetle's head and a smaller one arching up to meet it. When females are ready to mate (they spend most of their time below ground, feeding on plant material and possibly keeping out of the way of males), they waft out an attractive pheromone scent that brings males flying in. This is when the horns clash. The biggest beetles—the heaviest and longest—are those that have eaten the best food and so, possibly, have the best potential to father offspring. But they must prove themselves to onlooking females. Dueling males first threaten each other with head bobbing. Then the headbutting, levering, and tossing begins, and the loser is finally prized off his perch. The bigger a male is, the bigger his horns, the stronger his muscles and gripping legs, and the more likely he is to win. But bigger isn't always better. In the case of some horned scarab beetles, males that have invested in huge horns turn out to have small genitalia.

Strongest animal

NAME **rhinoceros beetle** *Xylotrupes gideon*

LOCATION Asia

ABILITY lifting possibly up to 850 times its own weight

Best surfer

Bottlenose dolphins are very social mammals, usually traveling in pods of 2–15 individuals and, offshore, in herds of tens or even hundreds. They often hunt cooperatively and, like many social predators, are inquisitive and adventurous. They frequently investigate people in the water, play with flotsam and jetsam and with other animals, such as sea urchins, and with seaweed (throwing these "toys" around, seemingly for fun), and will go out of their way to ride in the bow waves in front of boats and ships.

It's hard not to be anthropomorphic, because they seem to take pleasure in these apparently frivolous activities and rarely do them for any obvious practical reason. Bow riding, for example, is often a distraction from feeding or from traveling in a different direction. So perhaps it is inevitable that they should enjoy surfing. Common bottlenose dolphins and their close relatives the Pacific bottlenose dolphins, which are found in the western North Pacific and around the Indian Ocean, can be seen bodysurfing huge breakers close to the shore. Sometimes they are in the company of human surfers, racing in toward the beach and then swimming back through the surf and spray to wait for the next big wave. Given their other games and antics, it is almost impossible to come up with an explanation for the activity other than that they do it for the sheer thrill of the ride, just like we do.

NAME   **common bottlenose dolphin**
Tursiops truncatus
LOCATION   temperate and tropical waters worldwide
ABILITY   surfing and "bow riding"—probably just for fun

Earliest-rising songster

Bird-watchers have known for a long time that the thrush family—especially the robin, blackbird, song thrush, and American robin—are among the earliest daytime songsters. In surveys done in the United Kingdom of early avian choristers, robins and blackbirds inevitably come out on top, with the robin winning overall. Robins are also more likely to sing at night, when they are sometimes stimulated by streetlights, although night singing can be dangerous if owls or other predators are around. Males tend to do most of the singing: a male robin sings to defend his territory and to attract a mate. His singing decreases once he has a suitable partner, but he continues a certain amount of singing to keep her and to arouse her for mating prior to egg laying.

Dawn is the time when weather conditions are most favorable for song transmission, but weather is not the key factor in encouraging singing. It's a combination of daylight and male hormones. As daylight lengthens during the spring and early summer, there is a decrease in the production of melatonin (produced in the brain at night), and this triggers an enlargement of the song area of the brain.

Early songsters tend to have (proportionately) larger eyes than later-rising songsters. This helps them see better in low light: they usually start feeding earlier and can keep a watch out for predators. The male robin spends most of the day foraging and, if his energy levels are sufficiently high, he sings a little bit in the evening. But he tries to keep enough energy in reserve, after a relatively short night's sleep, to serenade full volume once again as the day dawns.

NAME **European robin** *Erithacus rubecula*
LOCATION Europe, south to north Africa, east to Siberia
ABILITY singing before dawn, often earlier than any other bird

The aardvark is one of the world's strangest mammals. It doesn't have any close relatives that are alive today, though distant ones include elephants, manatees, and golden moles, and no animals behave quite like it. The word aardvark is Dutch for "earth pig" and, being plump and hairless, it makes a juicy meal. A nocturnal lifestyle, combined with a good sense of smell and radarlike ears, helps it evade some predators.

But its chief means of defense is burrowing. It has long, chisel-like claws at the end of enormously powerful, muscular legs, and these make it possibly the fastest-burrowing animal in the world. Reputedly, it can dig a burrow in soft earth faster than two men can with spades. The burrows that it digs as daytime sleeping dens also act as escape holes and are long enough—often more than 33 ft. (10m) or so—to dissuade a predator, such as a hyena, from trying to dig out the aardvark or its babies.

The claws are also designed for digging termites and ants out of hard-baked earth. In fact, the aardvark's whole body is devised for "anting." At night it zigzags across the savanna with its snout down, and as soon as it detects a termite nest, it starts digging. Nose hairs filter out dust, and its 17-in. (45-cm) long, saliva-wet tongue extracts termites and ants at great speeds. Its termite-grinding teeth grow continuously, but they lack roots or enamel. A life of digging and dirt seems to work, though, since aardvarks are found in most parts of Africa where ants and termites are abundant.

Fastest digger

NAME	**aardvark** *Orycteropus afer*
LOCATION	African savanna, grassland, and open woodland
ABILITY	digging for a living

Longest toes

NAME **northern jacana** *Jacana spinosa*
LOCATION Mexico and Central America
SIZE 5.5 in. (14cm)—the longest toes relative to body length

The northern jacana (one of eight jacana species that are found in the tropics around the world) is 9 in. (23cm) tall and has an astonishing toe-span—4.5 in. (11.5cm) wide and 5.5 in. (14cm) long. That would be like a 6 ft. (1.8m) man having feet that were 3 ft. (0.9m) wide and 3.6 ft. (1.1m) long.

The jacana is sometimes called "the Jesus Christ bird" because of its apparent ability to walk on water. It can't exactly do that, of course, but it can walk on floating vegetation—broad stands of water hyacinth, water lilies, water lettuce—by spreading its weight across a wide area. This is where it spends most of its life, free of competition from birds and mammals that don't have such special feet.

The other distinction that the jacana is known for is its sexual "role reversal." A female usually has four mates, all occupying different territories. When it's time to breed, she copulates with all of them, again and again and in rapid succession, and when she lays her eggs—a clutch of four, normally—she leaves them with one of the males and rushes off to mate again. Even though there's only one chance in four that any particular chick will be his, the male left with the eggs dutifully incubates them and, when the chicks hatch, he takes care of them. He defends them, too, dashing back and forth on his pontoon feet at any sign of a threat.

Fastest swimmer

NAME **sailfish** *Istiophorus platypterus*
LOCATION warm waters worldwide
SPEED up to 68 mph (109kph)

It is notoriously difficult to measure the speed of a fish, and as no one has yet set up an open-water fish-racing track, we must rely on fishermen's estimates. The sailfish's predatory behavior and its body structure indicate a good capacity for speed. Like the nose of a jet, its rapierlike bill has been shown to cause what's called "low-resistance flow." And it is undoubtedly fast, recorded as reeling out 300 ft. (91m) of a fisherman's line in three seconds—faster than a sprinting cheetah (although estimates based on charging leaps out of the water aren't necessarily the same as top *swimming* speeds).

Close behind the sailfish in the speed-ratings are other open-water hunters: in order they are; swordfish, marlin, wahoo, yellowfin tuna, and bluefin tuna. One of the secrets of the sailfish's speed, along with that of these other fast predators, lies in its musculature. It has huge amounts of white muscle (great for acceleration but not for stamina), that are boosted along the flanks by blocks of red muscle (which need more oxygen but are good for sustained fast swimming). Most of the heat produced by the red muscle fibers is retained by a special network of blood vessels, making the blood warmer than the surrounding water. It can also channel warm blood to the brain and eyes, helping it spot and chase prey in colder water and at lower depths.

The exact function of its huge dorsal fin, the "sail," remains to be discovered, but it is thought to assist in fast-turning maneuvers, increase the fish's profile when it's rounding up prey, act as a sail when the fish is at the surface, and warm the blood when it is exposed to the Sun.

The water strider is a remarkable insect. It uses the surface tension of still water, such as on ponds, almost as if it were a thin layer of ice. Surface tension is not something that big, heavy animals notice much, but what happens is this: below the surface molecules of water are attracted to other molecules of water in all directions—up, down, and sideways. At the surface, though, the molecules cannot go upward any farther, and so it's all sideways and down. This creates what is basically a physical film that is firm enough to support something very light.

The strider's long legs also have microscopic hairs that trap air and make the legs more hydrophobic, or water-repellent, and the more hydrophobic something is, the heavier it can be and still be supported by the surface tension. The strider's legs don't get wet at all. As the insect moves across the water at speeds of up to 30 in. (75cm) per second, it makes little indentations, or dimples, in the surface tension and then uses the dimples as though they were the blades of oars—in fact, it rows across the water. And it only runs (or rows) on four of its six legs. The front two are very short and sensitive. They can tell when something breaks the surface, and if that something is small and keeps moving, the strider will dart over, grab it, and eat it.

Best water walker

NAME	**water strider** *Gerris* species
LOCATION	tropical and temperate zones worldwide
ABILITY	walking on water as if it were dry ground

Fastest long-distance runner

NAME	**pronghorn**, or **pronghorn antelope**, *Antilocapra americana*
LOCATION	plains of western North America and northern Mexico
SPEED	up to 55 mph (88.5kph) for 0.5 miles (0.8km)

The pronghorn is unique. It is neither an antelope nor a deer but is in a family of its own. It is built for both speed and endurance and, although the cheetah holds the speed record and can run faster, no other animal can run as fast over a sustained distance. The pronghorn has, for example, been recorded cruising at 42 mph (67kph) for one mile (1.6km), which would get it a speeding ticket in a residential area.

But why such speed? Scientists believe that, in prehistoric times, it was probably hunted not only by plains wolves, which have stamina, but also by a now-extinct predator with both stamina and speed—probably a super-cheetah. And, of course, there is nowhere to hide on the wide-open prairies.

Recent research has also revealed some of the pronghorn's physiological secrets. For a start, it has powerful lower torso muscles and extremely long, lightweight legs. When it is in full gallop, the front legs push all the way forward and the hind legs all the way back, allowing a long, airborne stride. Compared to most other mammals, its heart, lungs, and trachea are larger, and its blood is unusually rich in hemoglobin—which means that more oxygen can be delivered to the muscles in less time. It also has huge, protruding eyes that give it wide peripheral vision for spotting predators on the plains and binocular vision, which is critical for speed running. Alas, the pronghorn couldn't outrun the guns of European settlers, who by the early 1900s had reduced its numbers from possibly 40 million to just 10,000–20,000.

Most elastic tongue

At first look, a chameleon shouldn't be able to do what it does. It's spectacular enough that in only ten seconds the lizard can become an entirely different color, but the way that it uses its tongue makes even that seem mundane.

X-ray film and high-speed video footage have revealed that a chameleon's tongue on its way to an insect starts relatively slowly but then accelerates to 20 ft. (6m) per second in just 20 milliseconds. This is a faster acceleration than sheer muscular force could ever manage. Then when the tip of the tongue reaches the target, which can be more than two of the chameleon's body lengths away, it can attach to prey as heavy as 15 percent of the chameleon's own weight (large ones can grab birds or lizards) and, with nothing on the tongue stickier than viscous saliva, it can haul the prey back quickly and easily.

How? First, the shot: it's been discovered that between its tongue bone and tongue muscle a chameleon has some elastic collagen tissue that the muscle stretches before the tongue is released, the same way that a bowstring is stretched to shoot an arrow. Second, the grab: there's another muscle at the tip of the tongue that contracts in the instant before the prey is hit, converting the tip from convex to concave and making a powerful suction cup. Third, the retrieval: more tongue muscles and special filaments allow for "supercontraction," like an accordion slammed shut. And it all happens in a little more than one second.

NAME **chameleons** Chamaeleonidae family
LOCATION mostly Africa and Madagascar, but also the Middle East, southern Europe, and Asia
ABILITY capturing prey with a bowstring, suction cup, and accordion tongue

In the Middle East they run at 25 mph (40kph), jump 6 ft. (2m) in the air, and lay their eggs in camels' bellies. They're huge, and even more than that they're venomous. In Mexico they're called matevenados—deer-killers—and they're deadlier than scorpions. In South Africa they actually chase people.

Well, none of those stories is actually true, but you need only to see the frenzied way that a solifugid searches for and consumes its prey to understand how such tales have arisen. An eight-legged arachnid with a pair of pedipalps (appendages on each side of the mouth) that look like two more legs, it eats insects, spiders, and scorpions and sometimes stretches to small mammals or reptiles, depending on its size, from the smallest fraction of an inch to 6 in. (15cm) long. At night, when most solifugids are active, they zigzag across the ground at ferocious speeds, grabbing anything they come across and then stopping to crunch it with what may be the most powerful jaws, relative to overall size, in the animal kingdom.

Solifugids are insatiable. Toward the end of a good night's hunting, they can be so bloated that they can hardly move, but they'll still eat anything they can catch. They're not venomous, although a bite from a large one can draw blood from a human. As for chasing people, it's the rare diurnal ones that gave rise to that story. All they're doing is trying to cool off by staying in a human's shadow—which, nevertheless, must be incredibly disconcerting.

Scariest scamperers

NAME	**solifugids**, **wind spiders**, **wind scorpions**, or **sun scorpions** Solifugidae family
LOCATION	arid regions worldwide
ABILITY	frantic killing and eating

Best glider

NAME   **Japanese giant flying squirrel**
Petaurista leucogenys

LOCATION   Japan and S E Asia

ABILITY   flying—almost

Flying squirrels don't really fly. They don't have wings that they flap to give themselves thrust. But they do glide. Along with five other kinds of mammals—and certain lizards and snakes—they glide between trees to save energy and to avoid the dangers found while scurrying along the ground.

The best of the gliding animals (aside from windsurfing birds) would be the one whose glide is the longest, most agile, and comes the closest to flight. A typical glider goes close enough in a straight line from the tree it was in to the tree it's aiming for, floating on the outspread membranes between its forelegs and hind legs. But the giant flying squirrel is more agile than that. By changing the position of its legs, it can alter its direction and perform impressive fast turns and banks. It can even ride the warm air currents rising from the valleys in its native mountain home, like a soaring bird.

Its close relative, the red giant flying squirrel, is also an outstanding glider, and the two species can travel farther than any other nonflying glider—more than 361 ft. (110m). To travel such long distances, and to get up speed, they simply hold their legs close to their bodies and fall. They hurtle toward the ground at a heart-stopping rate and then open their membranes—and glide.

Every August and September monarch butterflies in North America receive a mysterious message from their genes. They stop their routines, either check the position of the Sun or sense the Earth's magnetic field (nobody knows for sure), and start flapping and gliding their way south. By November they're in the mountains of central Mexico— millions of them—blanketing groves of fir trees in about 20 different locations. It's the world's most massive insect migration: virtually all the monarch butterflies east of the Rocky Mountains participate.

Once they've settled on their trees, they don't move out again until February or March, when they snap out of their torpor, mate, and fly north. As they reach the southern US, the females seek out milkweed plants (nothing else will do) and lay eggs on them. Then all the butterflies die. When the eggs hatch, the caterpillars dine on milkweed, form their chrysalides, hatch and, as new butterflies, head farther north. Then two or three generations later (the summer butterfly generations live shorter lives than the overwintering ones), it's August again, and they get the genetic signal to migrate. So instead of becoming sexually mature and mating—as their parents, grandparents, and great-grandparents did—the monarch butterflies head for Mexico.

Sadly, recent studies indicate that the numbers of monarch butterflies sleeping for the winter in the fir forests are decreasing. The cause is probably a combination of illegal logging in their Mexican sanctuaries and the use of pesticides to kill milkweed plants in North America, and conservationists are now desperately trying to find ways to safeguard the world's biggest sleepover.

Biggest sleepover

NAME	**monarch butterfly** *Danaus plexippus*
LOCATION	North America
NUMBERS	millions overwinter together in Mexico

A cheetah actually spends most of its time resting from the heat, hiding from other big cats, or sitting high up looking for prey. But the action, when it happens, is sudden. After stalking its prey as closely as possible, it sprints from almost a standing start. Film of one chase showed the cheetah reaching 56 mph (80kph) in only three seconds. But the official record was set by a cheetah in Kenya that clocked an average 64 mph (103kph) over 660 ft. (201m).

At its top speed, a cheetah's incredibly flexible spine gives it a stride length that is about twice that of a racehorse's—so long that all four feet are off the ground for more than half the distance. Other design features include tough, gripping pads, nonretractable claws like running shoe spikes, very long, thin, springing legs, a rotary gallop that allows it to make sudden changes in direction, and a long tail to help balance those moves.

A cheetah doesn't always run at its top speed, however, and it usually gives up after 60 seconds. The average chase is no farther than 656–984 ft. (200–300m) and no more than 20 seconds, after which the animal, panting frantically, has to rest for at least 20 minutes to cool off and to let the build-up of lactic acid from its super-speed muscles dissipate. Unsurprisingly, prey animals, such as impalas and gazelles, have evolved to run fast, too—a Thomson's gazelle has been recorded at 58.5mph (94.2kph), fast enough to keep any cheetah on its toes.

Fastest sprinter

NAME	**cheetah** *Acinonyx jubatus*
LOCATION	grassland and semidesert regions of Africa, with possibly a few left in the Middle East and Asia
SPEED	up to 64 mph (103kph)

Longest migration

NAME **arctic tern** *Sterna paradisaea*
DESTINATIONS Arctic and Antarctic
DISTANCE 18,600 miles (30,000 km) round-trip

One look at the arctic tern tells you that it's an aerodynamic master and an excellent diver. Its body is slim and streamlined, its tail and wings are long and pointed to minimize the air resistance, its wings are elbowed for plunge diving, and its bill spearlike. The terns that breed during the northern summer in the high Arctic and "winter" during the southern summer in the Antarctic do a round-trip of at least 18,600 miles (30,000km) measured as a straight line. Many do even more, depending on their starting points and the weather conditions, preferring to stay close to the continental coastlines rather than fly cross-country.

The longest distance recorded for any single bird is held by a ringed arctic tern that left Finland on or around August 15, 1996. After flying via the English Channel and down the west coasts of Europe and Africa, it probably turned east at the Cape of Good Hope and crossed the Indian Ocean to arrive in Victoria, Australia, on January 24. That's more than 16,000 miles (25,760km)—about 100 miles (160km) a day. Undoubtedly it refueled along the route, as unlike land-based migrants, a tern is designed for sea fishing and can therefore fly offshore for the entire journey without needing to carry extra fuel and weight in the form of fat.

The breeding and wintering destinations are both worth the effort. In the high Arctic insect numbers are huge, benefiting chicks as well as adults, and fish are plentiful; the Antarctic, too, has huge numbers of fish and crustaceans. Both also provide almost perpetual daylight, allowing this high-performance bird nonstop, year-round fishing.

Biggest hunting army

NAME   **army ants** and **driver ants** *Eciton* and *Dorylus* species respectively

LOCATION   Central and South America (army ants); Africa (driver ants)

ABILITY   form raiding parties that may number up to 500,000 individuals

Many carnivorous animals hunt cooperatively. But driver ants of Africa and army ants of Central and South America—which live in colonies of up to 20 million—are the all-time record holders. With hundreds of thousands working together on a raid, they actually blur the line between related individuals working together and a single organism with many working parts. But however you look at them, when army or driver ants are on the march, they are formidable.

They move relatively slowly, and so larger animals—reptiles or mammals—can usually get out of their way. However, the sheer number of African driver ants, which have cutting "jaws" and form long, wide columns, can sometimes kill chickens and tethered domestic animals. But insects and other invertebrates, such as spiders and scorpions, are not as fortunate. Both army and driver ants sense potential prey mostly by the vibrations as it flees, and both operate essentially using the same division of labor. There's a great mass of workers, protected on their flanks by larger "soldiers" armed with powerful weapons: army ants have tissue-dissolving stings, and driver ants have powerful bites.

Army ant workers are the frontline raiders that do most of the killing, but soldiers guard them and the captured prey, and they are often stationed behind in order to catch any prey that has escaped the main attack. The booty is dismembered and carried by teams of porters back to the bivouac, the queen's mobile chamber, which is made out of the living bodies of workers and contains the colony's hungry larvae.

A sloth has two modes of being: not quite asleep and asleep. It can sleep for up to 20 hours per day: the longest life span that has been recorded for a sloth is 30 years, which means that the animal spent 25 years asleep. A sloth sleeps hanging upside down from a tree branch, which is also what it does when it's awake. The difference is that, when it's awake, it pulls leaves off the tree incredibly slowly and eats them incredibly slowly. Then it moves along the branch at a speed that's been worked out at 0.3 mph (0.5kph) or less than 45 ft. (14m) per minute.

On a really eventful day it descends to the ground and very slowly and awkwardly—since this is the only time it's ever upright—makes its way to the next tree. Sometimes the next tree is on the other side of a river or marsh, in which case the sloth swims to it, using a type of dog paddle that's strong and rather graceful compared to its walking but is, of course, very slow.

A sloth's metabolism is much lower than other mammals, and in the morning it gets up to speed by sunning itself. It also digests slowly and only defecates once a week: for this event it slowly comes down from its tree, slowly digs a hole, and deposits one third of its body weight (including urine) into it. Even the feces, which are hard and dry, decompose at around one tenth of the rate of other animals' feces.

Sleepiest animal

NAME	**brown-throated three-toed sloth** *Bradypus variegatus*
LOCATION	Central America and tropical South America
ABILITY	doing as little as possible

Extreme

Growth

Longest fangs · Greatest number of teeth · Fastest-growing plant · Smallest fish
Most gargantuan of growths · Heaviest land animal · Oldest living animal
Smallest reptile · Longest whiskers · Oldest surviving fish · Largest wingspan
Most elastic animal · Longest snake · Tallest living thing · Biggest brain
Largest superorganism · Longest tongue · Most numerous animal · Biggest bird
Largest animal ever · Longest weapons · Oldest living clone · Longest hair
Hairiest animal · Biggest mouth · Most feathers · Tallest animal · Most impressive
tooth · Largest eyes · Heaviest living thing · Fattest carnivore · Longest fin
Largest flower · Smallest amphibian · Most elongated fish · Baggiest animal
Oldest leaves · Rarest animal · Tiniest mammal · Oldest surviving seed plant
Largest reptile · Heaviest tree dweller · Flattest animal · Biggest canopy

Longest fangs

NAME   **gaboon viper** *Bitis gabonica*

LOCATION   west Africa

SIZE   up to 2 in. (5cm) long

With a maximum known length of 7.2 ft. (2.2m), the gaboon viper is the largest of the three African snakes that are in the group known as the giant vipers (the other two being the puff adder and the rhinoceros viper). It is one of the world's top-ten most venomous snakes and has the distinction of delivering the highest average yield of venom in each bite (actually, it shares this distinction with the king cobra of southern Asia—the world's longest venomous snake). It typically injects 350–600mg of venom and, since 60mg is considered a fatal dose for a human, a total yield of gaboon viper venom could theoretically kill six to ten people.

When it comes to the length of its fangs, the viper beats the cobra by no less than 1.5 in. (3.5cm). This, of course, means that the gaboon viper bites deeper than any other venomous snake. Why it should need to isn't clear—although it is able to swallow much larger animals, it mostly eats lizards and frogs. It seems unlikely that the long fangs are a defensive adaptation, because it isn't a particularly irritable snake and only rarely bites in defense. Perhaps the answer is simple: it is just a big snake, and so it has proportionately long fangs. But then why are the king cobra's fangs so short? Unlike the gaboon viper, whose fangs collapse backward when the snake's mouth is shut, the cobra's fangs are attached firmly in place. To put it simply, if they were any longer, they would puncture its lower jaw.

Greatest number of teeth

NAME   **whale shark** *Rhincodon typus*

LOCATION   tropical and warm temperate seas

NUMBER   more than 300 rows with hundreds of teeth in each

Exactly which animal has the greatest number of teeth is a bit of a conundrum. It depends on how you define teeth, how often they are replaced, and how long the animal lives. Mammals have teeth like the type we think of (enamel-covered, set into the jaw, and replaced only once in a lifetime), and the mammal with the biggest mouthful is probably the spinner dolphin, with up to 272 teeth. Crocodiles can have around 60 teeth, but these are replaced as many as 40 times—so that could be 2,400 teeth in a lifetime. However, snails and slugs have even more, if you call them teeth. They have a radula—a retractable, tonguelike file under their head—with rows of up to 27,000 microscopic teeth made of chitin, which are replaced as they wear down.

Shark teeth are specialized scales that are loosely embedded in a fibrous mass that is known as the tooth bed. In a conveyor-belt process they are continually shed and replaced by new ones. The shark with the greatest number of teeth is probably the whale shark, which is surprising since it is mainly a giant filter feeder. It has several thousand tiny, hooked teeth, each one around 0.08–0.12 in. (2–3mm) long, arranged in 11–12 rows on each jaw. These are probably replaced at least twice a year, and so, given that a whale shark could live as long as a human, it really does deserve the title of toothiest animal. But what it actually uses its battalion of teeth for is still a mystery.

Bamboos are strange plants. For a start, they are giant, woody grasses. Most of the 1,250 or so species do all their growing early in life. Once it is mature, a bamboo doesn't grow any taller, no matter how long it lives (and some survive for more than 100 years), preferring instead to send up more shoots. This means a stand of clumping bamboo, while not getting any taller, can become impenetrably thick.

The flowering is eccentric, too. Many species only flower once in their lives, when they are aged between 7 and 120, and then die. And that means that every single plant of a particular species may set seed at exactly the same time and die at the same time. (This is a special problem for giant pandas, which eat virtually nothing but bamboo and face general famine every 30–80 years, when the local bamboo species flowers.)

Bamboo is also hugely important to us (there are more than 1,500 documented uses for it), and up to 40 percent of the world's population depends on it. As for tortoiseshell bamboo it invests a huge amount of energy in seed production but may survive flowering, and it is widely grown as a crop. It's one of the tallest giant bamboos and probably the fastest growing. One shoot is recorded as putting on over three feet (1m) of growth in one single day—that's 1.6 in. (4cm) per hour—and another grew 65 ft 6 in. (20m) in eight weeks. It really is grass that you could watch grow.

Fastest-growing plant

NAME	**tortoiseshell bamboo**, or **moso**, *Phyllostachys edulis*
LOCATION	China and in cultivation worldwide
ABILITY	literally growing centimeters (even inches) per hour

Smallest fish

Not only is this the smallest of all fish, it's also the smallest and lightest vertebrate (animal with a backbone) yet to be discovered. It beats the previous record holder, the Philippine Lake Buhi goby, by 0.02 in. (0.5mm), although females are slightly larger. It is half the size of the other two known infantfish but with a deeper (more "stout") body. It is strange, though, not only because of its size but also because it's paedomorphic—meaning that adults resemble juveniles, hence the name infantfish. In addition, it has no teeth, scales, pelvic fins, or pigment (except in its eyes). It has large eyes, though no one knows why, as it hasn't yet been observed in the wild. Given that it has a large mouth but no teeth, it probably eats plankton.

Having fast-tracked the time that is needed to reach maturity by following a paedomorphic lifestyle and probably living for just a couple of months, this transparent, tadpolelike fish can reproduce fast and therefore might evolve fast. This could be a useful ability since it lives in protected coral lagoons that are comparatively free from temperature increases and storm waves—an environment that could change dramatically if global warming has its predicted effect.

Australian biologists speculate that the stout infantfish may be abundant and a significant part of a food chain, although only six individuals have been caught. The fact that it went undiscovered in a well-studied area of the world until recently, only being named in 2004, shows just how many other marine species have yet to be discovered, some of them possibly even smaller than this small fry.

NAME **stout infantfish** *Schindleria brevipinguis*
LOCATION Great Barrier Reef, Australia
SIZE only 0.2–0.3 in. (6.5–8.4mm) long

All of the records for humongous fungi are held by underground clone colonies of a species of honey fungus. The success of honey fungus lies with its "shoestrings," or rhizomorphs—parallel hyphae (rootlike extensions), covered in a tough, dark rind. These shoestrings travel great distances in search of food—vulnerable trees or deadwood—and pass back nutrients to the main mass of hyphae, called a thallus. They penetrate the bark of a living tree—often a young one—and siphon off water and nutrients from the sapwood. If the tree doesn't defend itself well, the rhizomorphs spread to the roots and effectively strangle the tree, extracting nutrients from the dying wood. Over the years new colonies form, and they are all clones of the original organism.

The largest-known mass of honey fungus, identified by DNA fingerprinting, is an *Armillaria ostoyae* clone in Oregon, three miles (5km) across in some places and estimated to be at least 2,400 years old (possibly twice that age). In the fall it grows great masses of "fruiting bodies," although the individual mushrooms are small by comparison with those of long-lived fruiting bodies of species such as the artist's fungus *Ganoderma applanatum*. Even larger honey fungus clones may exist in the huge conifer forests of Eurasia (*Armillana* especially likes certain conifers). Although the fungus might seem monstrous, it is part of a forest recycling system, creating spaces for new plants to grow and adding organic matter to the soil that provides nutrients for the trees.

Most gargantuan of growths

NAME   **honey fungus**, or **shoestring root rot**, *Armillaria ostoyae*
LOCATION Malheur National Forest in the Blue Mountains, Oregon, U.S.
SIZE   covers more than 2,198 acres (890 hectares)

The largest elephant that has ever been recorded was a bull elephant shot in Angola in 1974 (before the increase in poaching that decimated elephant populations). It weighed more than 12 tons—the equivalent of 178 men—and had a standing shoulder height of 13 ft. (3.96m) and a trunk-to-tail length of 35 ft. (10.7m). Another bull elephant, shot in Damaraland, Namibia, in 1978, had a standing height of 13 ft. 10 in. (4.2m), but the so-called desert elephants that live there have proportionately longer legs.

The average male African bush elephant is around 9 ft. 10 in.–12 ft. (3–3.7m), and females are smaller. Even smaller is the forest elephant at 6 ft. 7 in.–9 ft. 10 in. (2–3m), although it's still a heavyweight at 2–4.4 tons. Asian elephants (or Indian elephants), are closer in weight and height to the forest elephant, although slightly taller, the highest point being the top of the head rather than the shoulder. The Asian elephant record holder is probably a Nepalese bull estimated to be 12 ft. (3.7m) at the shoulder. Only white rhinos and hippos come close to such mammoths in terms of weight.

Unsurprisingly, elephants hold other records, including the largest appetite of all land animals, consuming 165–330 lbs. (75–150kg) of vegetable matter per day (double this amount in the case of large bull elephants). Despite an enormous stomach and up to around 62 ft. (19m) of intestines, their digestion is relatively inefficient, to the advantage of the animals who feed off its feces and the plants whose seeds it helpfully disperses in packets of fertilizer.

Heaviest land animal

NAME **African bush elephant** *Loxodonta africana*
LOCATION Africa, south of the Sahara
SIZE 3.9–6.9 tons

Oldest living animal

NAME **cold-seep deep-sea tube worm**
Lamellibrachia luymesi
LOCATION Gulf of Mexico
AGE at least 250 years

Tube worms are best known as a group of animals that, with the aid of chemical-converting bacteria, thrive in pitch darkness around hydrothermal vents, where hot gases and fluids spew out of the seabed and nourish forms of life that are not found anywhere else on Earth. Some of these vents can be as hot as 750°F (400°C), and the tube worms around them grow fast—more than three feet (1m) per year (see p.64).

But not all deep-sea vents are that hot, and in fact, not all of them are hot at all. There are also cold hydrocarbon seeps, for example, that release their fluid slowly and steadily and at the same cold temperature as the deep seawater. They support tube worms, too. Some of these tube worms can measure more than 6.6 ft. (2m) and appear to have taken a very long time to grow to such a length.

One year, deep in the Gulf of Mexico, scientists covered some of the tube worms' white outer casings with blue stain and returned the next year to find less than 0.4 in. (1cm) of new white growth peeping through. From that—and taking into account the faster growth rate of the young tube worms and the fact that they did not even collect the biggest and oldest animals—they calculated that an individual's possible life span is at least 250 years. Species such as giant tortoises and bowhead whales are believed to reach ages of at least 150 and 200 years respectively. However, the only other animals that might really rival tube worms are colonial ones such as corals, although with corals it's hard to distinguish one animal from another.

Smallest reptile

NAME **Jaragua gecko** *Sphaerodactylus ariasae*
LOCATION Beata Island, Dominican Republic (Hispaniola)
SIZE only 0.5–0.7 in. (1.4–1.8cm) from the tip of the snout to the base of the tail

If you want to look for new species that are either very small or very large, islands are where you are most likely to find them. And what is a better place for islands than the Caribbean? That's where the two record holders for the world's smallest reptile have been discovered. The first is the Virgin Gorda gecko *Sphaerodactylus parthenopion*, found in 1964 on the island of Gorda in the British Virgin Islands. Then in 1998 the Jaragua gecko was discovered on the island of Beata, part of the Dominican Republic's Jaragua National Park (after which it was named). Both the Virgin Gorda gecko and the Jaragua gecko average only 0.6 in. (1.6cm) from the tips of their snouts to the base of their tails.

The advantage to being small means that it's easier to hide and you don't need as much to eat, making it easier to survive on an island when food is more limited. But being very, very small means having a large body surface area relative to body volume, which puts both types of geckos at the risk of dehydration through evaporation. The Jaragua gecko survives by scurrying in and under moist leaf litter in remnant forests on the island, eating tiny insects, spiders, and mites. In fact, scientists think that it has probably taken the niche that, on the mainland, would normally be filled by a spider. Although it's not technically endangered, it may soon be classified as such, because a lot of the forest where it lives has been logged—barely 10 percent of the original forest remains in the West Indies.

Longest whiskers

A whisker may be just a rod of dead skin cells, but it functions as a sophisticated antenna. At its base is a blood-filled follicle, and when the whisker moves, it stimulates nerve cells. Depending on an animal's lifestyle, whiskers (vibrissae) may be any length and thickness and are usually arranged in several areas on the face: on the cheeks, on or around the nose, and above the eyes.

The animals that have the longest whiskers and accomplish the most with them are usually those that are active in darkness or in low-light conditions, which includes the marine mammals. Some seal whiskers may each be serviced by more than 1,000 nerve cells (compared to around 250 for a rat). Effectively, they function as eyes and fingers. They may pick up messages not only about texture, shape, and size, but also about movement and water pressure—because in the water any movement leaves a wake, or "footprint," of whatever has passed by.

Antarctic fur seals do a lot of their hunting at night, for krill and squid, and in the Antarctic winter night is virtually constant. Their whiskers are arranged in precise patterns on their cheeks, and when a fur seal is hunting, its whiskers can point forward, feeling ahead, like those of a cat when it's hunting. A bull Antarctic fur seal has the longest whiskers of all. But why, no one knows. Maybe he uses them to express his inner emotions or maybe he just needs them to look good—signaling that he's a master hunter when he's parading his prowess on the breeding grounds.

NAME	**Antarctic fur seal** *Arctocephalus gazella*
LOCATION	Antarctic waters
SIZE	13.7–19.5 in. (35cm–50cm) long

Coelacanths are not only ancient, but they are also the only surviving members of their group and among the oldest of all vertebrates (animals with backbones). They were thought to have become extinct 65 million years ago, until 1938, when a living one was hauled up in a fishing net off South Africa. Surviving populations have been discovered since throughout the western Indian Ocean (including the Comoros archipelago, Madagascar, Kenya, Tanzania, Mozambique, and South Africa), and a second, related species (*Latimeria menadoensis*) was recently discovered thousands of miles away, off Sulawesi, Indonesia.

Differing little from their ancestors, living coelacanths are up to 6 ft. 6 in. (2m) long, with seven lobed, paddlelike fins—distant relatives of the animal that took the first steps on land around 350 million years ago. No coelacanth has been observed alive at the surface, probably because the fish live in cold, well-oxygenated water below 328 ft. (100m) and die at the warm surface, where there's less dissolved oxygen.

A coelacanth has the largest-known fish egg, roughly the size of a grapefruit and weighing up to 12 oz. (350g). The female carries as many as 26 of these at once for a gestation period of around 13 months—a very heavy load. It also appears to have an electrosensory organ in its snout, possibly to help it hunt in the dark, and has been observed gathering in caves and doing "headstands." Biologists are now using special diving techniques in an attempt to discover more about the social life of this mysterious living fossil.

Oldest surviving fish

NAME	**coelacanths** *Latimeria chalumnae* and *L. menadoensis*
LOCATION	western Indian and Pacific oceans
AGE	400 million years

Largest wingspan

NAME **wandering albatross** *Diomedea exulans*

LOCATION southern oceans

SIZE up to 11 ft. 1 in. (3.4m) wide

The wandering albatross has the longest of all wings. They are narrow, flat, light, and designed for soaring and high-speed gliding in the strong, steady winds of open oceans. It rides the waves and weather fronts and can cruise at 34 mph (55kph) for hundreds of miles and reach 55 mph (88kph) over short distances. It may not breed until it is six to eight years old and can live to be 50 or more. It builds up a vast ocean map, learning the best foraging areas and remembering the location of the remote island of its birth, normally returning there every other year to display or nest—the only time it alights on land.

The smaller, lighter females can maintain maximum glide speeds using lighter winds, and so they forage farther north than the larger, heavier males, who are more energy efficient in the windier subantarctic. This allows a pair to scour a huge combined area, covering up to 8,060 miles (13,000km) on a single trip, and cuts down on competition for food.

Large tube nostrils enable them to locate feeding areas using scent cues—chemicals released over upwelling zones and seamounts. An albatross's favorite food is deep-water squid, which it may catch by sitting on the sea at night, waiting for them to come up to the surface. But it's also a scavenger, often following ships for garbage and fishing scraps. And here lies its downfall. Huge numbers of wandering albatross drown every year scavenging bait from the hooks of longline fishing fleets, threatening the species with extinction.

Most elastic animal

NAME **nemertean**, or **ribbon worm**, *Lineus longissimus*

LOCATION northern Atlantic

SIZE can be more than 98 ft. 4 in. (30m) long when stretched out

This is not only the most elastic creature in the world, but it's also one of the most bizarre predators. It's a nemertean worm. Like many of its relatives, it is basically one long, muscular tube, with a brain but no heart. Although the more usual length of this ribbon worm is around 16-49 ft. (5-15m), individuals of 98 ft. (30m) have been recorded. The extreme length record may, however, result from the fact that the worm can elongate its body considerably—very useful for getting under rocks and tying itself up in a knot. Like most nemertean worms, it can regenerate lost or damaged body parts if it gets stretched to breaking point.

Despite its appearance, it's a hunter, searching out marine creatures such as other worms, snails, crustaceans, and even fish. It can see, using a row of up to 40 small eyespots on each side of its snout, but it probably tracks prey by chemical trails rather than by sight, since it is often found squirming around in dark places such as in muddy gravel and under boulders. The ribbon worm's secret weapon is a powerful, long, muscular, tonguelike proboscis that it shoots out in order to lasso prey. Its body has a purplish iridescence, making it seem even more unearthly. And it's an animal to be wary of. If you pick one up, you'll find yourself covered in thick, sticky mucus with a strange smell—a smell that could mean it's just released some of the potent tetrodotoxins that it uses for defense.

Giant snake stories are abundant—due to the wild imaginations of many early explorers, the difficulties in estimating or measuring the length of live animals that refuse to stay still, and the fact that large snakeskins can be deliberately stretched without causing very much noticeable distortion. Most modern herpetologists have a healthy skepticism about any snake that is claimed to be longer than 30 ft. (9m).

The most extraordinary tales concern the South American anaconda. It rarely grows longer than 20 ft. 6 in. (6m), yet it is probably the subject of more exaggerated claims about size than any other animal. Anacondas spend most of their time in the water and so, admittedly, they can support an enormous weight (the heaviest snake ever recorded was an anaconda weighing 500 lbs. (227kg). But the famous claim by Lt. Col. Percy Fawcett in 1907 of one that was 62 ft. (18.9m) long has since been shown to be a big exaggeration.

However, female reticulated pythons do reach 20 ft. (6m) relatively often, and since length comes with age, and constrictors tend to live a long time, exceptional lengths are feasible. In fact, the longest snake that was properly recorded is a female reticulated python, supposedly measuring 32 ft. 9 in. (10m), which was shot on the Indonesian island of Sulawesi in 1912.

A large female is sufficiently strong, long, and expandable enough to stop the blood flow and breathing of a large mammal and swallow it whole. In fact, there is at least one reliable record of a reticulated python with an adult human in its stomach. The top records for length, though, are still likely to remain those that were recorded in the past, as most snakes, persecuted the world over, rarely get to live to a ripe old age.

Longest snake

NAME **reticulated python** *Python reticulatus*

LOCATION Southeast Asia

SIZE 20 ft. (6m), and possibly up to 32 ft. 9 in. (10m)

Tallest living thing

NAME   **the Stratosphere Giant**, a California redwood, *Sequoia sempervirens*

LOCATION   Humboldt Redwoods State Park, California, U.S.

HEIGHT   370 ft. 2 in. (112.8m)

The Stratosphere Giant may now be the tallest tree in the world—and the tallest living organism. But the tallest tree ever to be reliably reported—the Cornthwaite Tree, a huge mountain ash in Victoria, Australia—was measured after it had been felled in 1855 as 374 ft. (114m). The tallest mountain ash today, growing in Tasmania, is relatively short at 318 ft. 3 in. (97m). This means that all of the tallest trees now are California redwoods, and there are around two dozen that are taller than 360 ft. (110m).

Redwoods have been extensively logged, and so it's very likely that there have been taller ones than these in the past. But how much taller? According to calculations made in 2004—taking into account factors such as the pull of gravity and the limits of water friction—it's possible for a California redwood to grow to between 400–426 ft. (122m–130m).

So what goes on 50 or 60 stories up a tree? For one thing, these trees sprout whole new trees, known as reiterated trunks. One redwood that was studied had a whole forest in its crown—209 reiterated trunks. Most of them were fairly small, but the largest was 8.5 ft. (2.6m) in diameter and 131 ft. (40m) tall. Deep soil had also accumulated up there—in crotches and on large branches—and growing out of it were ferns, shrubs, and other trees, although not necessarily redwoods. There were also plenty of insects and earthworms, mollusks, and even a sizable population of salamanders.

Identifying the biggest brain of all is a bit of a brainteaser, depending on whether you mean just the largest or the biggest, meaning the brainiest. And is brain size actually linked to intelligence? You really need to be a brain scientist to be exact about the matter.

Big animals, of course, have big brains. The brain of the sperm whale weighs in at a massive 17.2 lbs. (7.8kg), while the biggest land animal, the African bush elephant (see p.170), has the biggest terrestrial brain—up to 11.9 lbs. (5.4kg). By comparison, a human brain is relatively small. But relative to body size, it's actually bigger. So does that mean that humans are at the top of the intelligence chart? This is where it gets complicated. If we take intelligence to mean the ability to reason—and the degree of complexity of the neocortex (an area of a mammalian brain that is involved in thought and communication) being an indication of that ability—then humans win outright. But the role of the neocortex is still being investigated, and as far as many experts are concerned, our brain should still be regarded as simply a bigger version of a primate brain.

However, some scientists believe that our brain has undergone more rapid evolution than that of any other animal, fueled by the development of highly complex social structures and behavior. For instance, we've developed the ability to deceive others—and even ourselves. And despite what evidence about the state of the world might suggest, the human brain may, according to one theory, still be evolving at great speeds. It makes you think.

Biggest brain

NAME **human** *Homo sapiens*

SIZE 2.9 lbs. (1.3kg) on average

ABILITY thinking

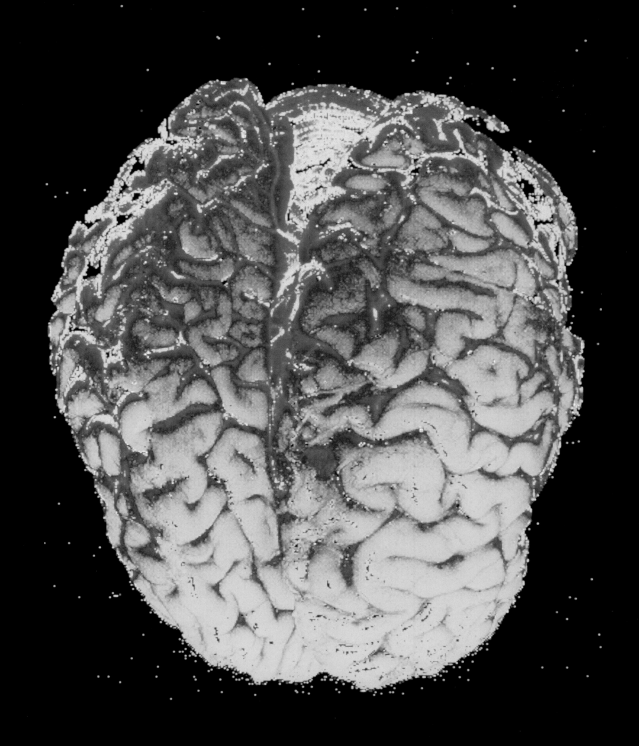

Ants are arguably the most successful animals on Earth, estimated to comprise 15 percent of the planet's total biomass. It's all due to cooperation: individuals take roles and work as if they are cells in one superorganism. Members of a colony are the offspring of the queen, and in evolutionary terms it makes sense to sacrifice individuality in order to help their relatives. Some ants, however, take such togetherness to the extreme.

In the 1900s a tiny, brown, innocuous ant stowed away among plant material that was being exported on boats from South America to the U.S. and as far afield as South Africa and Australia. These original Argentine colonizers were few in number, but their tiny genetic stock has led to big things. In new, warm lands, without their South American parasites and with no significant restrictions on numbers other than

availability of water, they have multiplied, outdoing native ants by sheer mass. The largest supercolony of all is composed of millions of genetically related and interconnected nests and stretches at least 3,720 miles (6,000km) from northern Italy to northern Spain.

The ants' success lies in fecundity (nests have numerous queens, and so they breed fast) and peace. Unlike nests of native populations, they don't bother fighting each other, which leaves more time for food gathering, reproduction, and defense. This unusual social organization is unlikely to be watered down by new Argentine ant arrivals, because any ants that are not recognized as close relatives are attacked. And so the supercolonies, from California to Australia, are likely to grow and grow, with new ones forecast for Asia.

Largest superorganism

NAME	**Argentine ant** *Linepithema humile*
LOCATION	has spread to six continents
ABILITY	forming supercolonies of billions of ants

Longest tongue

NAME   **Morgan's sphinx moth** *Xanthopan morganii praedicta*

LOCATION   Madagascar

SIZE   a proboscis 12–14 in. (30–35cm) long—the longest-known "tongue" relative to body size

This is probably the most famous tongue of all. It was first brought to the attention of the scientific world through the imagination of Charles Darwin, the great 19th-century natural philosopher and father of evolution. In 1862 he examined a specimen of the comet orchid that grows in the forest canopy in Madagascar. It has large, waxy, white, star-shaped flowers that produce a powerful, sweet scent at night. What fascinated Darwin was that its nectar was at the bottom of a spur that was around 12 in. (30cm) long, and he was convinced that the structures and pollination devices of orchids had developed in tandem with their insect pollinators.

Knowing that white flowers that were scented at night attracted moths, he wrote in 1877: "In Madagascar there must be moths with proboscides [the 'tongues' through which they drink nectar] capable of extension to a length of between ten and eleven inches!" As the flower offered insects no landing pad, it was likely to be a hovering hawk moth. Darwin was ridiculed for his suggestion. But in 1903 Morgan's Sphinx Moth was discovered, with a proboscis to match the length of the spur.

For many years the relationship between the two species in the wild was unconfirmed, but the moth has recently been observed visiting the orchid flowers and carrying away pollen. However, a further mystery remains. The comet orchid's close relative, *Angraecum longicalcar*, has a spur measuring 16 in. (40cm), indicating that there is another moth yet to be discovered with an even longer tongue.

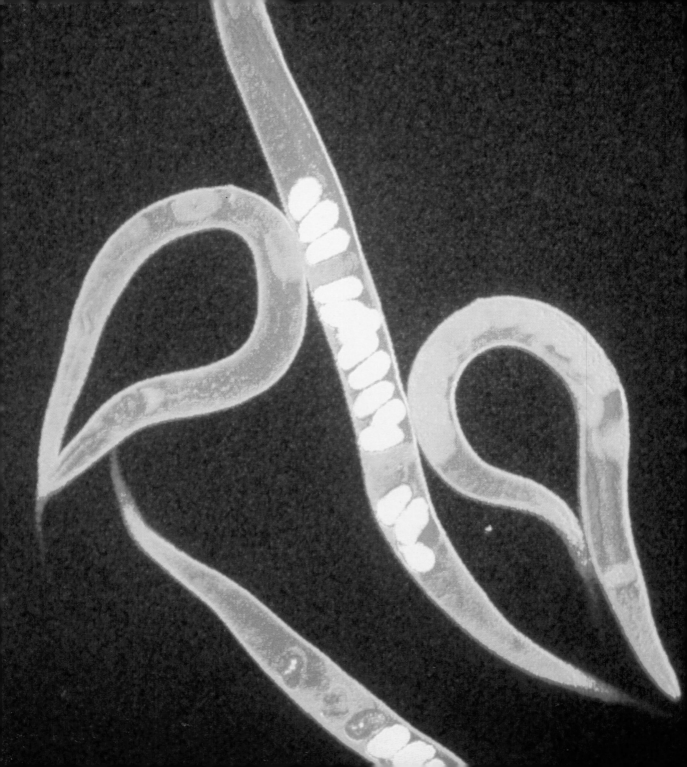

Most numerous animal

These are the most numerous multicellular animals, outnumbered only by single-celled bacteria. What they demonstrate is that, to be a multicellular animal, you don't really need a lot: an outer surface, muscle, a nerve or two, a mouth, a gut, a way of excreting, and a way of reproducing. That's about all most nematodes have, and with those basics, they're as compact as an animal can be and have become the most successful animal on Earth—in fact, they probably comprise four out of five of all animals.

If you pick up a handful of soil, you'll also be holding thousands of worm-shaped nematodes. In an acre of rich farmland there can be three billion nematodes. There are nematodes at the poles and at the deepest point on the seabed. If conditions ever get too hot or cold or dry, they can suspend animation, shutting down all their body functions until things improve. There are also more different species of nematodes than in any other phylum, except for the arthropods (insects, spiders, and their relatives): 20,000 nematode species have been described, but that's barely a start. Most of the nematodes that scientists know best are microscopic parasites—on humans or on other animals or plants—but there probably is not a creature on the planet that isn't parasitized by one. It's as parasites—especially intestinal ones—that nematodes become large, and the larger the host, the larger they will grow. One found in a sperm whale measured 26 ft. (8m) in length.

NAME   **nematodes** in the phylum Nemata
LOCATION   everywhere—in the sea, on land, and in you
NUMBER   billions of billions—probably four out of every five animals

This is a multiple record beater among birds. It's not only the tallest and the heaviest, but it is also the fastest runner with the largest eyes and the biggest egg. Being a herbivore (plant eater) and very large means that it needs to spend most of its time eating, and it travels many miles per day to get its fill. When it flees from predators, its huge thighs are capable of powering it at up to 45 mph (72kph), aided by antelope-like, two-toed feet (other birds have four toes).

Its scientific name, *camelus,* probably refers to the fact that it can survive with little water. It makes use of moisture in succulents and other plants and, aided by its long trachea (windpipe), saves water by cooling inhaled air so that it contains less moisture when it is exhaled. Huge wings also act as sunshades and fans, and a comparatively featherless neck and legs can lose heat fast.

Although it can defend itself with its powerful legs, it's a gentle creature—unlike Stirton's thunderbird, the largest bird that ever lived. This flightless Australian giant was almost ten feet (3m) tall and weighed 1,105 lbs. (500kg). Nicknamed the "giant demon duck of doom," based on its duckish ancestory and huge bill, this was almost certainly a bird with a taste for bone cracking. Bird-headed, kangaroo-like giant creatures feature in prehistoric Aboriginal paintings and might just be pictures of this nightmare creature, although it probably became extinct around 26,000 years ago.

Biggest bird

NAME	**ostrich** *Struthio camelus*
LOCATION	drier regions of Africa, south of the Sahara
SIZE	up to 9 ft. (2.8m) tall and 353 lbs. (160kg) in weight

The blue whale has the largest set of superlatives in the animal world, including the heaviest body, the loudest voice, the biggest appetite, and the smallest prey (in relation to body size). It's also one of the most mysterious animals—despite its phenomenal size, we know surprisingly little about many aspects of its life. It grows to an average of 78.7–88.5 ft. (24–27m) long, with a maximum-recorded, world-record length of more than 110 ft. (33m) and a record-breaking weight of 187 tons.

It seems surprising that it can be so big when it eats tiny, shrimplike creatures called krill, but a blue whale gulps down around three tons or more of these highly nutritious crustaceans per day. With the loudest voice in the animal kingdom, its low-frequency sounds can literally travel hundreds or even thousands of miles, although no one knows whether such powerful vocalizations are used to communicate over great distances or for underwater imaging to help with long-distance navigation.

The blue whale's size and speed saved it in the days of sailboat whaling, but one of the grimmest blue whale facts is that more than 350,000 were killed following mechanization of the whaling fleets during the last century. Almost all populations were drastically reduced in size, some by 99 percent. Now the hunting of blue whales is banned, but only the population that spends the summers off California seems to be thriving. Everywhere else in the world, there is serious concern about the future of these remarkable giants.

Largest animal ever

NAME	**blue whale** *Balaenoptera musculus*
LOCATION	oceanwide
SIZE	the largest animal ever to have lived

These relations of the jellyfish are so primitive—one first appeared 650 million years ago—that they are not even considered single animals but, instead, colonies of five types of organisms: float, sensory, stinging, digestive, and reproductive. The tentacles come in two sizes—short ones that are massed under the float and one or several (depending on the species) extremely long, trailing ones for deeper fishing. Sometimes these tentacles will stretch outward on the surface and sometimes in opposite directions—and when they do, they make the man-of-war, at least temporarily, longer than a blue whale, and thus the longest animal on the planet (putting aside the "elastic" ribbonworm—see p.182).

Most of the time, though, these long tentacles hang down. When they are touched or stimulated by chemicals, such as animal proteins, their tiny stinging cells (nematocysts) release barbed threads that puncture skin and inject nerve poison. A small fish that bumps into a tentacle is immediately paralyzed and held, while muscles at the top contract to haul up the catch. Then the fish is engulfed and bathed in enzymes, which break it down into nutrients for distribution to the other members of the colony.

The notoriety of the Portuguese man-of-war, as far as humans are concerned, lies in its randomness. It hunts randomly and goes wherever the winds and ocean currents take it. That can often be inshore waters or a beach. When it's beached, of course, it dies, but the nematocysts continue working, inflicting excruciating stings on thousands of people every year.

Longest weapons

NAME	**Portuguese man-of-war**, *Physalia* species
LOCATION	tropical and subtropical open oceans
LENGTH	stinging tentacles up to 116 ft. (35m) long

Oldest living clone

NAME **Kings holly**, or **Kings lomatia** *Lomatia tasmanica*

LOCATION just one gully in southwest Tasmania

AGE more than 43,600 years old

This remarkable shrub was discovered around 70 years ago by the self-taught naturalist Deny King, while he was panning for tin on a mountain range in southwest Tasmania. But that area was later burned, and the plant was never seen again. Then in 1965 King found another population in a cool, rain forest site on a range farther east. It was confirmed as a new species, a member of the Proteaceae family (with its closest relative in Chile, hinting at the time when Australia and South America were joined together), and was named in King's honor.

The plant is sterile, being genetically triploid (having three pairs of chromosomes instead of two). Since it cannot reproduce sexually, its attractive red flowers never set seed, and it can only regenerate by shooting from root suckers. This also means that all 600 or so plants in the population are genetically identical clones.

Even more extraordinary, fossil leaves of the species that were collected in the same region and are visually identical to living leaves have been radiocarbon dated as at least 43,600 years old. Because of this and because it reproduces clonally, biologists believe that the living Kings holly plant and the fossil are, for all practical purposes, the same plant. This makes the species the oldest-known clone, beating the creosote bush (at 11,700 years old) and the previous record holder, the North American huckleberry (at around 12,000 years old). It was alive when *Homo sapiens* and Neanderthal man existed together, but tragically, this ancient survivor is now threatened with extinction.

Longest hair

NAME **musk ox** *Ovibos moschatus*

LOCATION Arctic Alaska, Canada, Greenland, Europe

LENGTH up to 2 ft. 9 in. (90cm) long

That famous naked ape, *Homo sapiens*, can grow hair longer than 16 ft. 5 in. (5m), the world record being 16 ft. 11 in. (5.15m). But human hair is more for sexual display than for warmth, which is the usual function of hair. So maybe the record should really be for the longest known body-warming hair. And the musk ox coat certainly is warm—several times warmer than the coats of any of its relatives, sheep and goats.

Musk oxen, in fact, have two coats—the long, shaggy one that hangs almost down to the ground and a fine, short, very warm undercoat, as much as 12 in. (30cm) thick in places, called (in Inuit) the "qiviut." The two coats together get musk oxen through winters that are as cold as -94˚F (-70˚C). Such superinsulation also allows them to lower their metabolism by around 20 percent, to help them survive on much less food during the long, dark, freezing-cold winters.

If the musk ox looks like a tundra mammal from the last ice age—something that might have gone the way of the woolly mammoth—that's because it is. It shared the tundra with the mammoth and, along with a few other mammals, it survived the new, warm era by adapting to climate changes, sticking close to the rim of the Arctic Ocean and not migrating. Much later the arrival of men with guns almost finished it off, but it was rescued by "eleventh-hour" conservation efforts and has now been reintroduced back into some of its old haunts.

Hairiest animal

NAME   **sea otter** *Enhydra lutris*
LOCATION   coastal Pacific waters
NUMBER   almost one million hairs per square inch (up to 394,000 per square centimeter)

A sea otter spends most of its time on its back in the sea, resting, grooming, or feeding—cracking open shellfish on an anvil that it places on its tummy. It therefore needs to be both buoyant and ultra-warm. To maintain its body temperature, it burns calories at almost three times the rate that we do, and so every day it has to eat at least one quarter of its body weight in fish and shellfish. Blubber, used by most sea mammals for insulation, would make a sea otter too heavy to float. Instead it invests in the thickest coat of any animal, comprising up to one billion hairs.

A fine, downy undercoat traps air for insulation and is overlaid by long guard hairs measuring 1.4 in. (3.5cm) and providing a covering that is so dense that water doesn't penetrate it. To keep the coat in perfect condition, the otter spends up to half of the day grooming: combing out dirt, straightening and aligning the hairs, blowing air into the undercoat, and spreading waterproof oil.

A baby sea otter is born in a flotation jacket—a buoyant undercoat—and can't dive for a couple of months until this has been replaced by adult fur. However, the sea otter's luxurious coat has almost been its downfall. Fur hunters brought the species close to extinction, and although it is protected today, oil spills pose a new hazard. Once an otter's coat is contaminated by this oil, it loses its insulation, and the otter dies of hypothermia or from ingesting the oil as it tries to clean itself.

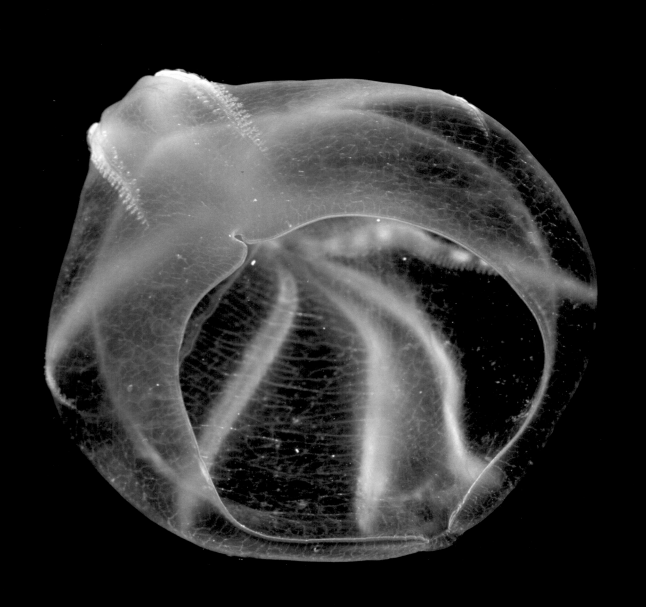

The biggest mouths are found in the sea. The widest mouth belongs to the whale shark (see p.162), and the biggest food scooper of all is the huge, pleated throat of the blue whale (see p.198). Both animals are effectively filter feeders, eating relatively tiny prey. The most dramatic large mouths, though, are those that engulf large animals whole and probably belong to deep-sea creatures such as the gulper or the loosejaw—elongated animals with huge (relative to size), hinged jaws.

But when it comes to mouth relative to body size, they cannot compare to the ctenophores (beroid comb jellies). A beroe is, effectively, an enormous stomach that is surrounded by a thin, muscular, gelatinous wall, which opens up wide by means of a huge mouth. It swims by beating eight rows of ciliary comb plates along its sides, has light sensors but no eyes, and can probably "smell" its prey.

While it's cruising, a beroe's lips are tightly held shut with adhesive strips. As soon as a beroe bumps into prey—anything from other comb jellies to fish—its giant nerve net instantly triggers the lips to part, and muscles snap the mouth open at incredible speed, enabling it to engulf the creature whole. If the prey is too large to swallow, the beroe bites off a huge chunk with the thousands of tiny sharp-tipped "teeth" (macrocilia), that line its lips and may extend into its stomach. The lips then reseal the "mouth-of-all-mouths," and the bloated beroe swims away to digest its enormous meal.

Biggest mouth

NAME	**ctenophores**, or **beroid comb jellies** *Beroe* species
LOCATION	most oceans
SIZE	2–11.7 in. (5–30cm) long (or wide)—all stomach and mouth

Most feathers

NAME **tundra swan**, or **whistling swan** *Cygnus columbianus*
LOCATION North America
NUMBER 25,216 feathers

Back in 1933 a group of bird-watchers decided to count the feathers on a dead tundra swan. The amazing thing—aside from the fact that they really did spend hours counting—is that they found 20,177 feathers on the swan's head and neck alone. In fact, many birds may have more than one third of their feathers here (principally"under down"), perhaps to protect their brains against temperature extremes when they fly very high.

Tundra swans pair for life, and so they don't need to grow display plumage every year to attract new sexual partners. But they do feed on the water and so they preen constantly, using oil from a gland at the base of their tails to keep their plumage flexible, insulated, and waterproof. They also need to replace worn feathers. The adults molt their "primaries," or flight feathers, simultaneously from July to mid-August, while they are on their Arctic breeding grounds and before they start their long southbound migration in the fall. But the remainder of their thousands of feathers are gradually lost from the summer to the following spring. So, although you might find quill feathers at a nest, or a lakefront molting ground where nonbreeding birds gather, you will never find more than a tiny proportion of a tundra swan's 25,000 body feathers at any one site.

The bird that holds the record for the least number of feathers is supposedly a ruby-throated hummingbird, with 940 feathers. But relative to body size, this tiny species has many more feathers than a tundra swan, which is about 2,000 times larger in size.

Evolution often has a way of giving certain animals a monopoly on one aspect of their environment, and giraffes are a very visible example of that. No other large mammal, except for the occasional elephant, can browse in the higher reaches of the acacia trees that provide the basis of a giraffe's diet. Its sensitive and dexterous lips and long (18 in. (45cm), prehensile tongue are also designed for the task of carefully stripping out leaves from their thorny protection. In this way, a giraffe can get around 75 lbs. (34kg) of virtually uncontested, nutritious leaves per day. It also obtains enough moisture from its diet to survive without drinking. If there is an opportunity to drink, it has to spread its front legs out wide, get its head down, and rely on a complicated series of vascular valves to keep its blood from rushing down there, too.

Its height, size, and excellent vision mean that an adult giraffe is not an easy animal to ambush during the day, and other animals may even use giraffes as predator lookouts. Lions, however, are a threat, especially at night. Because after dark, a giraffe's height and vision are of limited use when it's lying down and ruminating (chewing unprocessed food that is stored in a special part of its stomach). However, giraffes are capable of giving a powerful kick. And if a giraffe should sense lions in time, it can gallop away, with a rolling, leggy gait that belies its speed: 31–37 mph (50–60kph).

Tallest animal

NAME	**giraffe** *Giraffa camelopardalis*
LOCATION	Africa, south of the Sahara
SIZE	males up to 18 ft. (5.5m) high

Nature is full of weapons—horns, antlers, and tusks among them—that are grown by males for fighting over or merely impressing females. But only one animal develops such a weapon from a single, long tooth. This is the narwhal. The male's tusk, which pierces its upper lip, is actually its upper left front tooth. Instead of staying embedded in the gum, as it does in females, it pushes out through the upper lip, twisting (always counterclockwise when it is viewed from the root) as it grows, into a lance with an average length of 6 ft. 7 in. (2m)—roughly half the length of the whale's body. Exceptional tusks can be as long as 9 ft. 10 in. (3m) and weigh up to 22 lbs. (10kg). They look similar to gnarled, twisted walking sticks.

The role of the tusk baffled scientists for years. Among many theories were that it was used for spearfishing, grubbing for food, and drilling through ice. In fact, males have been observed crossing tusks at the surface, and the presence of scars and wounds around the head region, combined with the high incidence of broken tusks, suggests that they are used for fighting and as displays of strength. The narwhal's tusk was thought to have been the horn of the legendary unicorn until early in the 1600s, and traders and alchemists conspired to keep the existence of the whale a secret while selling "unicorn horns" with magical healing properties for huge profits. Queen Elizabeth I paid £10,000 ($18,000) for one.

Most impressive tooth

NAME	**narwhal**, or **narwhale** *Monodon monoceros*
LOCATION	Arctic waters
LENGTH	a spiraling front tooth up to 9 ft. 10 in. (3m) long

These eyes belong to a formidable predator that hunts in the dark depths of the coldest oceans. Since few adult specimens havebeen examined so far, exactly how it uses its huge, protruding eyes is not known. But it can light up its surroundings literally by glowing, enabling it to focus its dinner-plate-sized eyes on fast-moving prey such as the large Patagonian toothfish—which the only intact specimen, an immature female, was snacking on when it was caught in 2003 by fishermen in the Ross Sea.

Large eyes go with the most massive body of any known invertebrate—even larger than the better-known giant squid. The Ross Sea specimen weighed 331 lbs. 5 oz. (150kg) and was, with its tentacles stretched out, almost 17 ft. 7 in. (5.4m), with a massive 8 ft. 2 in. (2.5m) mantle (body) and huge head. In fact, the species may grow up to 49 ft. (15m), with a mantle up to 13 ft. (4m). Unlike the giant squid, it has a physiology that suggests that it can move fast, a large muscular fin, and up to 25 razor-sharp and swiveling hooks on clubs at the ends of its two extended tentacles. It also has a huge, parrotlike beak, making it one of the most fearsome predators in the sea.

Largest eyes

NAME	**colossal squid** *Mesonychoteuthis hamiltoni* (the photo shows a model)
LOCATION	cold, deep oceans
SIZE	24 in. (60cm) or wider

This is one huge tree complex—a giant stand of genetically identical tree trunks connected by a common root system and weighing thousands of tons—called Pando, or "I spread," in Latin. Although the individual trees, or ramets, are comparatively short lived, there are at least 47,000 of them—all males—and the clone itself is at least 10,000 years old. It may even be much, much older. And although the ramets are comparatively slender and rarely get to be very tall, the area that the tree complex covers is at least 106 acres (43 hectares).

The quaking aspen can reproduce in a normal, sexual way, by producing seeds. But if the conditions are not good for seed germination, or if the aspen is damaged by fire or an avalanche, it opts for fast, vegetative reproduction, throwing out new suckers to replace fallen trees and continuing the spread of the organism. In fact, being partially fireproof, it thrives on periodic fires, which kill off competing trees.

A mature root system like that of the giant clone is also capable of putting out almost one quarter of a million shoots per acre (half a million per hectare), and since aspen shoots can grow three feet (1m) per season, it can outdo other trees. As a result, the quaking aspen successfully colonized North America in the wake of the last ice age and is now the most widely distributed tree on the continent—and the second only to the juniper as the most widespread tree in the world.

Heaviest living thing

NAME	**quaking aspen** *Populus tremuloides*
LOCATION	Wasatch Mountain Range, U.S.
SIZE	5,900 tons

Fattest carnivore

NAME **polar bear** *Ursus maritimus*
LOCATION Arctic and subarctic
ABILITY a female is capable of putting on four times her weight in one season

Polar bears are large. Their average weight is an impressive 442–1,326 lbs. (200–600kg), but there is an unsubstantiated record of a male that was allegedly shot in Alaska and weighed an incredible 2,210 lbs. (1,002kg). By comparison, the most heavily-built bears—grizzlies from the Gulf of Alaska called Kodiak bears—have a record weight of only 1,656 lbs. (751kg).

Pregnant female polar bears spend most of the fall and early winter asleep and most of their waking time searching for food (mostly fat-rich seals and their pups). By the end of a good spring they can sometimes have fat deposits amounting to more than 50 percent of their body weight. This makes them the record holders for the greatest fat intake of any adult land mammal. Females are smaller than males, yet their seasonal increase in weight is no less than four times greater. During their winter sleep a female polar bear has a lower metabolic rate and reduced heart and breathing rates, and she doesn't eat, drink, urinate, or defecate. But her body is fueled by fat reserves, and her temperature rarely drops more than a few degrees below normal.

Unfortunately, climate change is causing the pack ice on which polar bears hunt to freeze later and thaw earlier. This makes hunting difficult and, consequently, the average weight of polar bears in some populations is decreasing. Scientists are worried that in the future, females may not be able to put on enough weight to survive both their winter sleep and the effort of rearing young cubs.

There are three known species of thresher sharks—the common, the bigeye, and the pelagic—and there is also some evidence of a fourth species from Baja California, in Mexico, which is yet to be described and named. But the common thresher shark is the largest and has the longest tail, or caudal, fin—possibly up to ten feet (3m) long in the biggest individuals.

All thresher sharks have the same basic shape and the exceptionally long upper lobe of the scythelike tail is almost a tail in itself. It's not the longest fin in the world—that probably belongs to the humpback whale (see p.32), which has enormous pectoral fins, or flippers, that can be more than 16 ft. (5m) long. But it's the longest relative to body size.

The thresher shark is believed to use its fin as a whip, by swimming around a school of small fish or squid in ever-decreasing circles and then using it to stun or kill the frightened animals with a hefty whack. Thresher sharks have been seen killing seabirds in this way, but few of these sharks have been observed feeding underwater, and so there is actually little direct evidence to support this intriguing theory. However, in some parts of the world killer whales (see p.48) use a similar technique, and anglers have reported catching thresher sharks on live baits—not hooked in the mouth, but in the tail. Like sharks all over the world, thresher sharks are fished for the commercial value of their fins, which are used to make expensive shark fin soup.

Longest fin

NAME	**common thresher shark** *Alopias vulpinus*
LOCATION	tropical and temperate oceans
SIZE	up to 20 ft. (6m) long, almost half of which is its tail fin

Largest flower

NAME   **Rafflesia** *Rafflesia arnoldii*

LOCATION   Borneo and Sumatra, Indonesia

SIZE   up to 3 ft. (0.9m) in diameter, 25 lbs. (11kg)

The flower is all you'll ever see of a Rafflesia, because the rest of this intriguing plant is comprised of just threads inside of a tropical vine. It is a slow-growing parasite, totally dependent on *Tetastigma* vines, from which it draws nutrients.

It is very rare for a Rafflesia bud to emerge from its host on the forest floor, and about nine months pass before it swells to the point when it literally bursts open. Its five red fleshy lobes (sepals) curl back to release a putrid stench like rotting meat and reveal the huge, open-topped dome that houses the reproductive parts (which can be either female or male). The smell attracts carrion-loving beetles and flies. As the pollinators don't fly far, flowers need to be nearby and their flowering synchronized so that pollen can be moved from male to female flowers within just a few days. How this happens is a mystery. Another mystery is how any of its thousands of tiny seeds land on a vine, although it's possible that they get there via the dung of the tree shrews and squirrels that eat the fruit. Somehow, a seed penetrates the stem and starts growing. But it will be years before it buds.

Rafflesia's scientific name commemorates two famous 19th-century plant and animal collectors—Sir Stanford Raffles, the founder of Singapore, and the botanist Joseph Arnold. Sadly, as its host is disappearing with its shrinking rain forest home, Rafflesia itself is quickly becoming history.

Smallest amphibian

NAME   **Mount Iberia frog** *Eleutherodactylus iberia*
LOCATION   Cuba
SIZE   only 0.4 in. (1cm) long

If you are Brazilian, you'll claim the Brazilian gold frog to be the smallest. If you're Cuban, you'll make the claim for the Mount Iberia frog. Both average around 0.4 in. (1cm) long. But given that Cuba has several other close contenders for the title—including the Tetas de Julia frog, named after the mountains on which it was found (literally the "breasts of Julia" frog), and the more aptly-named yellow-striped frog—it seems fair to give the record to Cuba. It is certainly the place of tiny frogs. In fact, Cuba has one third of the Caribbean's amphibians, and an astonishing 94 percent of these are not found anywhere else in the world—although many are threatened by deforestation, exotic invaders, such as rats and cats, or mining.

The Mount Iberia frog was discovered in 1993 by the Cuban biologist Alberto Estrada on an expedition in search of the exceedingly rare ivory-billed woodpecker (he was probably the last person to glimpse the huge bird in Cuba in 1986, although it has since been rediscovered in the U.S. state of Arkansas). He located the frog by its "hissing" call (in the process of miniaturization, its "larynx" has become tiny and it has lost a few teeth). When he saw its metallic copper stripes and purple belly, he felt sure that it was a new species to science. Most biologists now consider it to be the world's tiniest tetrapod—meaning that it's the tiniest of all four-legged animals with a backbone.

Washed-up specimens of this serpentlike fish are reputed to be the origin of at least some sea monster sightings. Although this shimmering, ribbonlike creature has been seen swimming close to the surface, it probably lives in deeper water, and so most of its behavior and life history remains a mystery. It has no tail, but running the entire length of its body is a dorsal fin (described as bright red in near-surface and beached specimens) that undulates. The oarfish appears to use this fin for hovering, and it has been seen to swim vertically up and down in the water—highly unusual for a fish. Two long pelvic fins, each one reduced to a single ray and ending in a yellow and blue tassel, are held out like oars.

The few people who have seen an oarfish swimming say that it moves fast, and one person even describes it as curious. It has large eyes, reminiscent of deep-sea predators. Being toothless, it's perfectly harmless to humans, and some people believe that it's a filter feeder. However, it does have a comparatively large, protrusible mouth and is presumably capable of sucking in small shrimps, fish, and squid. It may be that the tassels on its rays and the long, antennae-like spines on its head are fishing lures or sensitive feelers, and so it may hang in wait for prey, its fine-scaled, silver body reflecting the blue of the sea and making it almost invisible. The oarfish must certainly be a puzzle to predators that are used to horizontal fish, especially if it decides to escape, vertically, tail first, at great speeds—a truly amazing sight.

Most elongated fish

NAME	**oarfish** *Regalecus glesne*
LOCATION	in temperate and tropical seas worldwide
SIZE	up to 50 ft. (15.2m) long

Baggiest animal

NAME	**giant Titicaca frog** *Telmatobius culeus*
LOCATION	Lake Titicaca, Andes (Peru/Bolivia)
ABILITY	breathing through its baggy skin

Large lakes cut off from other bodies of water are like isolated oceanic islands, where species can evolve independently from evolution that is happening anywhere else on the planet. One of the world's most isolated lakes is Titicaca in the Andes, which is only fed by snowmelt and rain—not by sizable rivers.

At an altitude of 12,516 ft. (3,815m), the 3,220 sq miles (8,340 sq km) lake is also one of the world's highests and as such it has a very low oxygen content. So the most important adaptation of the lake's aquatic animals (and they are almost all exclusive to Titicaca) is an exceptional ability to process what little oxygen there is. Among these is the giant Titicaca frog, one of the few amphibians that never needs to surface for air. Although it does have rudimentary lungs and can breathe air if it has to, but mostly it stays underwater, absorbing oxygen through its skin. And the more skin it has, the more oxygen it can absorb.

So, entirely as an aid to respiration, its skin hangs off it in great, flabby folds and adds up to more than twice as much as the frog would need if its skin were "skintight." It is, incidentally, among the world's largest frogs, with one having been recorded at 20 in. (50cm) and 2.2 lbs. (1kg). The larger the frog becomes, the proportionately more skin it needs in order to get enough oxygen. This means that the biggest Titicaca frogs are also the baggiest.

This plant is essentially just a pair of leaves on a short, bowl-like stem. Individuals get to be very, very old. Their straplike leaves grow and grow and don't drop off. But they do get thrashed by the wind, and so they are never longer than around 20 ft. (6m)—rather than the 656 ft. (200m) they could be—and they are sometimes horribly split and tangled.

Welwitschia usually grows close enough to the coast to make use of the fog that rolls in from the Atlantic Ocean at night. Moisture condenses on the leaves and is both channeled down to the root and taken in through the breathing pores, or stomates. Sometimes welwitschia conserves water using a special method of photosynthesis (the way that plants make food). The leaves open their stomates and take in carbon dioxide, not during the day, like most other plants do, but at night, when temperatures are coolest and they won't lose water through transpiration (evaporation through the stomates). They then store the carbon atoms as special acids until the Sun's up and they can photosynthesize—using the light to synthesize the carbon into carbohydrates.

Instead of flowers, welwitschia produces cones like other gymnosperms (conifers, ginkgoes, cycads, and seed ferns). These are either male cones with pollen or female cones that form the seeds. Both of them produce a sticky fluid—in the case of females, to capture pollen, and in the case of males, to attract insects to disperse pollen—a combination of characteristics of both flowering and nonflowering plants.

Oldest leaves

NAME	**welwitschia**, *Welwitschia mirabilis*
LOCATION	Namib Desert, southwest Africa
AGE	can live to be more than 1,500 years old

Rarest animal

This is the only remaining Pinta Island Galápagos giant tortoise—and you can't get much rarer than that. It has always been island wildlife that has suffered the most, because isolated islands with unique species were seen by ships' crews as convenient mid-ocean supply stops. To 19th-century sailors, the huge tortoises on Pinta and other Galápagos islands were very handy: each animal was worth many meals, and one could live for up to one year without being fed. Then goats were released on Pinta as a future food source, and they stripped the island of the tortoises' staple plants.

The last Pinta Island giant tortoise was seen in 1906—until George was found in 1971. He's around 80 years old now, residing at the Darwin Research Center on Santa Cruz Island. But George is a subspecies, and there are still giant tortoises on other islands. So what is the rarest whole species?

It could be the Yangtze river dolphin, which is so rare that only a handful of individuals (if any) survive. The last known Hawaiian po'ouli, a small brown and black bird, died in 2004, but it's not certain if others survive or not. And there was the 1984 case of the world's most surprised ornithologist: one night he was smacked in the head by a Fiji petrel, a bird that was last seen in 1855. Other petrels have been spotted since, but no one knows exactly how few remain. Sadly, the list of possibilities is almost endless.

NAME	**Lonesome George** *Geochelone nigra abingdoni*
LOCATION	originally from Pinta Island, Galápagos Islands now on Santa Cruz
STATUS	only one individual left

There is a lot of debate over which species is the record holder, but the mammal with the smallest head-to-body length is a bumblebee-sized southeast Asian bat, which is also one of the rarest bats in the world. It was only discovered in 1973 by Dr. Kitti Thonglongya (hence its alternative common name, which also refers to its piglike nose) in limestone caves in Thailand, and another population has since been discovered in Burma.

Because so few individuals have been weighed and measured, it's difficult to be exact about dimensions. But it would seem to be slightly heavier than the other contender for "smallest mammal," the widespread Savi's pygmy shrew, which weighs only 0.4–0.9 oz. (1.2–2.7g) and is 1.4–2.1 in. (3.6–5.3cm) long, not counting the tail (the bumblebee bat has no tail as such).

Being so small is a problem for both mammals, because they have large body surface areas relative to body volume, resulting in great heat (energy) loss. They therefore need to eat a lot frequently. Both are voracious hunters, the bat mostly catching mainly flies, and the shrew chasing anything small enough or slow enough to tackle. To aid their frantic lifestyles, they have huge and very fast-beating hearts and special fast-contracting muscles. But if it gets cold or food is scarce, both opt for the same survival option—they become torpid and wait until conditions improve.

Tiniest mammal

NAME **bumblebee bat**, or **Kitti's hog-nosed bat**, *Craseonycteris thonglongyai*

LOCATION Thailand and Burma

SIZE only 1.14–1.30 in. (2.9–3.3cm) long, weighing only 0.06–0.11 oz. (1.7–3g)

Oldest surviving seed plant

NAME   **ginkgo**, or **maidenhair tree**, *Ginkgo biloba*

LOCATION   originally throughout northern temperate zone; now in gardens and streets

AGE   the genus dates back 280 million years

This tree is the ultimate survivor. When Hiroshima, Japan, was bombed in August 1945, virtually every plant in the city was destroyed, including a particular ginkgo tree, which was burned and irradiated. In the spring of 1946 the remains of that tree produced one sprout. Today it's a perfectly healthy tree growing in the grounds of a temple 0.6 miles (1km) from the center of the blast.

Ginkgoes antedate many things—flowers, for instance. When ginkgoes first appeared on Earth 280 million years ago, there were no angiosperms—plants that produced flowers and encased their seeds in fruit. In fact, as a member of the Ginkgoaceae family—the last survivor, in fact—*Ginkgo biloba* is somewhere between the primeval cycads, conifers, and primitive ferns and the first flowering plants. This species has held on through Earth-shaking volcanic cataclysms, asteroid collisions, and the general environmental change that caused all the plants that were the ginkgo's early contemporaries to die out or evolve into something else.

Its real problems were ice ages. It was frozen out of North America seven million years ago and Europe three million years ago. But ice never reached parts of southeastern China, and that was where they enlisted the help of a very newly evolved species—*Homo sapiens*. Trees were found by the ancient Chinese and cultivated in their temple gardens. Now, although probably extinct in the wild, they thrive in cities all over the world because of their resistance to air pollution and disease.

This is the world's biggest reptile, weighing up to 2,646 lbs. (1,200kg). Male saltwater crocodiles are mature at around 10 ft. 6 in. (3.2m), and the smaller females at 7 ft. 3 in. (2.2m), but they continue to grow and may live to be one hundred years old. The longest authenticated record is of a male at just over 23 ft. (7m). But a fairly reliable record does exist from the 1950s of a beast of over 33 ft. (10m) that lived along the Segama River in Sabah, Borneo—its length was measured from its impression in the sand. These days, though, huge individuals that are longer than 20 ft. (6m) are rare.

A saltwater crocodile's eyes, ears, and nostrils are positioned on top of its enormous head, allowing it to lie in wait for prey, with the bulk of its body hidden below the surface of the water. Massive jaw muscles enable it to exert almost one ton of force—enough to crush and lock its interlocking teeth onto prey—and since it can remain underwater for hours, it can easily drown large mammals.

Being an indiscriminate feeder,taking anything from fish and birds to other crocodiles and mammals—a saltwater crocodile will even eat people. A male protecting his territory or a female guarding her young may also be aggressive if they are disturbed. And so in areas where people live and bathe in waters where there are crocodiles, attacks on humans are common. It's the reptile to treat with the greatest respect.

Largest reptile

NAME	**saltwater** or **estuarine crocodile** *Crocodylus porosus*
LOCATION	freshwater, but also brackish water and saltwater; from India to southeast Asia and Australasia
SIZE	up to a maximum of 33 ft. (10m) long

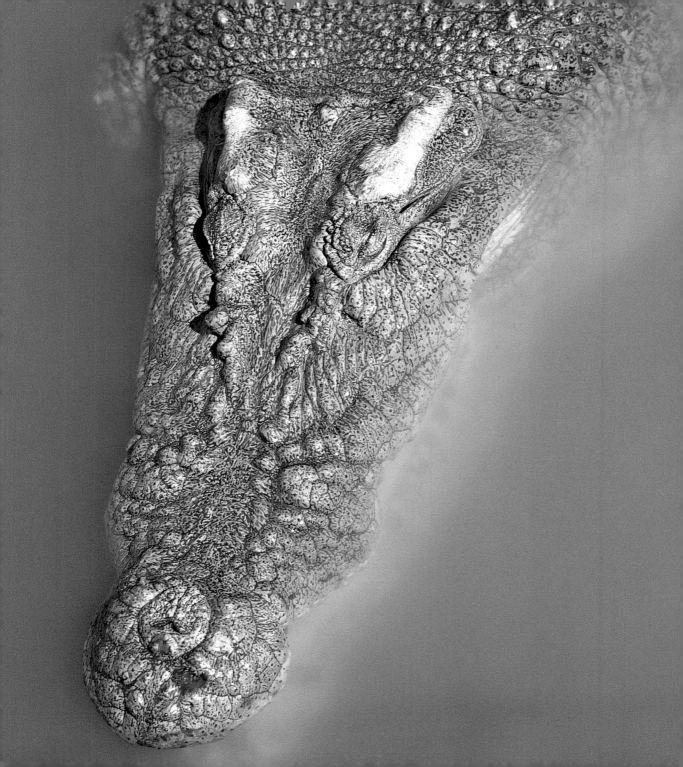

Heaviest tree dweller

NAME **orangutan** *Pongo* species
LOCATION Sumatra and Borneo, Indonesia
SIZE males can be up to 300 lbs. (35kg)

Life in the trees is something that's normally reserved for lighter mammals—partially because the heavier you are, the fewer branches there are that can support you, and partially because heavier mammals tend not to have the nimbleness that tree-top navigation usually requires. But primates evolved in the trees. Climbing, after all, is what hands are for, and recognizing fruit is the original reason for primates developing color vision.

However, the larger great apes have, in part, left the branches for the ground. Lesser apes, the gibbons, are superb tree climbers, but apes that are bigger—chimpanzees, bonobos (see p.299), and gorillas—have lost a lot of the knack. Pygmy chimps, or bonobos, tend to only spend part of their day aloft, although they sleep in tree nests. Common chimps sleep in tree nests, as well, and will climb for food, but spend much of their waking life on their feet and knuckles on the ground. Gorillas, the largest apes, even though they climb, in most parts of their range seem to prefer the forest floor.

There is one great ape, however, that spends most of its life in the trees: the orangutan. It is lighter than a gorilla but heavier than a human, and it seems conscious of its weight. It doesn't scamper like a monkey or swing like a gibbon. Instead, it moves slowly, deliberately, and skillfully with all four limbs, almost as if it were strolling through the foliage.

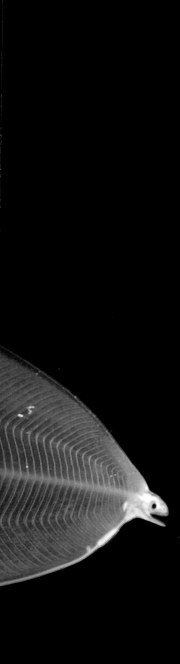

None of the flat, transparent animals in the oceans are quite as fascinating, or as long, as leptocephali. Eels as a group have one of the most eventful life cycles on the planet, and it all starts with a leptocephalus, the ancient Greek word for "slender head." This refers to the head on the larva of the European eel *Anguilla anguilla*—the best studied of all eels and a source of fascination since Aristotle declared that eels came from earthworms. However, it wasn't until 1893 that anyone made the connection between the European eel and what had been known as *Leptocephalus brevirostris*. And it's no wonder. Who would have thought that a 0.04–0.08 in. (1-2mm), transparently flat, slightly fish-shaped thing in the sea was the same animal as the familiar snaky inhabitant of European freshwater?

This is what biologists believe happens: at around 10–14 years of age adult eels get the overwhelming urge to swim downriver and across the Atlantic to the Sargasso Sea. How and why is a mystery. It's believed that they spawn in the Sargasso (but no one has seen them doing it) and die. The eggs hatch into leptocephali, which drift in the Gulf Stream for three years, finally reaching Europe. Now 1.8 in. (4.5cm), they metamorphose into elvers and migrate upriver. Years later, as adult eels, they will get the call to make the long, hard journey to the sea to produce another generation of extremely flat leptocephali.

Flattest animal

NAME	**leptocephali**, or **glass eels**, *Anguilla* species
LOCATION	Atlantic and Pacific oceans
SIZE	elongate and leaf thin

In A.D. 70, Pliny the Elder, one of the first great natural historians, wrote, "There is in India a tree whose property it is to plant itself. It spreads out mighty arms to the earth...." He was writing about adventitious roots, which are the secret of the banyan's reputation as the world's greatest shade-giving tree.

Like many of the almost 2,000 other species of fig trees, the banyan can send roots down from its branches. These create pillars that support the branches as they continue to spread. The carefully tended banyan in the Indian Botanic Garden, Calcutta, with its 2,800 prop roots, is the largest known banyan. All over southern Asia the trees are cared for and used as gathering places—as markets or schools or for village assemblies—and, indeed, the name comes from *banias*, or merchants, because that was where English traders did business with the locals.

Other fig trees that are famous for their adventitious roots are the "stranglers," which germinate from seeds in another tree's canopy and send roots down the trunk. Eventually, the fig tree crushes the host to death, leaving just a tall fig tree in its place. The success of fig trees is partially the result of the adaptability of their roots, but these aren't always adventitious. When growing like normal trees, they take the soil as it comes, thin, and the roots spread out deep, and they go straight down. The deepest roots ever recorded are those of a fig tree in South Africa, measuring 400 ft. (120m).

Biggest canopy

NAME	**banyan tree**, *Ficus benghalensis*
LOCATION	southern and southeast Asia
SIZE	the largest has a diameter of 1,378 ft. (420m)

Extreme

Families

Deadliest love life · Strangest incubation · Most agile embryo
Male with the most mates · Flashiest male · Greatest number of spores
Longest continuous incubation · Most artistic suitor · Most reproductive organs
Most extreme mating ritual · Strangest nuptial gift · Longest lasting love bite
Biggest seed · Greatest mass egg laying · Largest egg · Shortest mammal
gestation · Sexiest immaculate conception · Most fertile animal · Largest nest
Strangest society · Most upwardly-mobile gender change · Most extreme
reproductive act · Biggest group-spawning event · Most colorful male · Sexiest
animal · Most traumatic insemination · Sparkliest animal · Most imaginative use
of dung · Oldest juvenile · Most quarrelsome siblings · Greatest sex divide
Longest pregnancy · Strangest nesting material

Deadliest love life

NAME **agile antechinus** *Antechinus agilis*

LOCATION Australia

EVENT every male dies after mating from stress

This little insect-eating, marsupial mouse has a short but promiscuous life. Like all antechinus species, agile antechinus have a brief, two-week mating season when life becomes supercharged. Being secretive, nocturnal tree climbers, their mating behavior in the wild isn't well known, but they are being studied in captivity by biologists who are eager to understand the effects of reproductive stress.

In July or August males become flooded with testosterone and other hormones, and that's when the mating frenzy starts. But it's the smaller females who make the running. The males gather together in tree nests, where the females go to look for mates. The females seem to prefer dominant and therefore larger males, but they don't seem to be too fussy, since they mate with several different individuals. A male, too, mates several times. Yet ejaculation goes on for at least three hours, and he stays locked to his mate for up to 12 hours, to make sure that his sperm gets to her storage site first.

Alas, the surge of hormones and all of the effort is just too much for his immune system. If gastric ulcers and kidney failure from stress don't kill him, infections or parasites do, and he dies within days of his copulation—along with all other males in all the populations. However, some females live to mate a second year. They also have the upper hand when it comes to the gender of the next generation, as the sex ratio of their babies (which are kept in a true marsupial pouch) is often skewed toward females.

Strangest incubation

Both of these remarkable frogs are probably extinct. The last southern gastric-brooding frog, *Rheobatrachus silus*, was seen in 1981, and the last northern gastric-brooding frog, *R. vitellinus*, in 1985. Strangely, the northern species remained common until March 1985, but it was absent three months later and hasn't been seen since. Their extinction is a terrible loss—not just of two exceptional species but also of a whole unique method of incubation and brooding.

As their common name suggests, the frogs would incubate their eggs and then brood their young inside of the mother's stomach. They were the only animals in the world known to do such a thing. The female literally swallowed her fertilized eggs, which developed into tadpoles and then froglets inside of her stomach. She could not eat during the gestation period of six to seven weeks, and eventually she gave birth through her mouth, with one or two fully-formed froglets at a time hopping into the outside world. During this remarkable process the production of the digestive secretion, hydrochloric acid, actually stopped—effectively turning the stomach into a temporary womb.

In both species the brood size was 20–25, and the whole delivery took around a day and a half. After another four days the digestive tract returned to normal and the female resumed feeding. Why these frogs became extinct is still unclear, although timber harvesting may have been partially to blame. There have been intensive searches for them since, but to no avail.

NAME **gastric-brooding** or **platypus frogs**,
Rheobatrachus silus and *R. vitellinus*
LOCATION Queensland, Australia
ABILITY incubating babies in their stomachs

This little baby is a miraculous climber. It pops out from its mother's cloaca as a virtual embryo—blind and not much bigger than a broad bean—almost small enough for the mother not to notice that she has given birth. Then it has to journey on its own up to the pouch. It has only partially developed its forelegs and must drag itself up its mother's fur. Yet surprisingly, it can complete the journey to the pouch in record time—not much longer than about three minutes.

Once it has clamped onto a nipple, it resumes a type of embryo existence, fattening up on its mother's rich milk—the consistency of which changes as the baby matures and requires different nutrients. It will be as long as four months before the joey releases from the teat, six months before it first leaves the pouch, and eight months before it leaves for good. It continues to suckle for another four months by sticking its head into the pouch and, should danger threaten, will also use its mother's enormous pouch as an escape hole.

A female red kangaroo can, in fact, mate and produce a new embryo shortly after giving birth, but this embryo remains dormant until its sibling has left the pouch. It attaches to a vacant, short teat, while its sibling uses the nipple that it has always suckled from. Should anything happen to the joey before it's weaned, the embryo will immediately start maturing—an efficient insurance policy for hard times.

Most agile embryo

NAME **red kangaroo** *Macropus rufus*
LOCATION Australia
ABILITY climbing at least 4–6 in. (10–15cm) from its mother's cloaca to her pouch

Male mammals are capable of fathering many more offspring than females, and so, for some mammals, it makes sense for one male to collect as many females as he can, defend them from other males, and then sire as many offspring as possible. This is known as "polygyny." The mammal that has taken polygyny the furthest is the male southern elephant seal, a few of which mate with up to 100 females per season (although 40–50 is more usual).

Adult males spend most of the year at sea, then arrive on the breeding beaches in August. The biggest ones claim territories where they know that females will come ashore, in September and early October, to have the pups that they conceived a year earlier. When the females have given birth, they go into estrus, and the dominant males mate with as many as possible in order to have the maximum number of offspring the following year.

Of course a few males with lots of mates means lots of males with no mates at all—and for every breeding colony there's a huge adjacent bachelor colony full of disgruntled seals with little more to do than sneak over and try to mate with the females next-door. What distinguishes the southern elephant seal from the northern elephant seal in the North Pacific, which has a similar mating system, is that northern bachelors are more successful at sneaky mating, making their southern relatives the current record holders for the most polygynous mammal.

Male with the most mates

NAME **southern elephant seal** *Mirounga leonina*
LOCATION subantarctic islands and southern Argentina
ABILITY mating with more females than any other male mammal

Flashiest male

NAME   **morpho butterflies** *Morpho* species

LOCATION   Central and South American rain forests

ABILITY   producing vivid flashes with the most iridescent of living structures

Of all the butterflies, the iridescent blue morphos have always been the most sought after by collectors, and their wings have even been used for jewelry. The males' vivid color flashes, visible even from aircraft flying over the rain forest canopy, are not for wooing females, however. The iridescence is in fact for intimidating rivals and claiming territory.

Such intense color is produced not by pigments but by possibly the most complicated reflective structures in the world. The wing scales (the "powder" that rubs off a wing) are angled like roof tiles. Each scale supports another series of layered, almost transparent scales, which may be layered with further structures. The purpose is to reflect light not just upward but also outward. The structures are so precisely arranged that light rays of a particular wavelength are reflected back in the same but parallel direction, enhancing each other and therefore the reflected color. Result: extreme vivid color.

Being so flashy is dangerous, however (females, by comparison, are camouflage brown). So the males have another trick. When they fly, the movement of their wings up and down alters the angle of light striking the scales, and the color changes suddenly from bright blue to brown. The effect, enhanced by an undulating, chaotic flight and the exposure in the upward stroke of brown underwings, is that the butterfly appears and disappears. And as soon as it settles and closes its wings, exposing just the brown underside, it melts into the forest.

Most large fungi spread themselves by releasing into the air multitudes of microscopic spores from special fruiting bodies—mushrooms and toadstools—that grow aboveground. Some use animals, water, or even plants to help move the spores, but most rely on the wind to blow them away, sometimes over huge distances. The spores are usually released by the use of special turgid cells, which means that most fruiting bodies can't risk drying out. These mushrooms and toadstools therefore usually grow when it's damp.

However, the giant puffball doesn't rain spores from a multitude of downward-facing pores or gills in the usual way. Instead, it bears them internally, keeping them adequately humid, and releases them gradually. If conditions are moist enough, it matures in just one week or so, swelling to huge proportions—sometimes to more than three feet (1m). It then begins to split open, aided by the occasional animal knock and abrasion, over weeks or even months to release billions or trillions of spores into the wind. This isn't considered a very efficient method of dispersal, since most spores will neither travel far nor survive, which is why it produces so many of them. But then the job is done if just a few spores settle and "germinate" in favorable (nitrogen-rich) nearby fields. It's also a good thing that all the spores don't grow into puffballs. If they did, in just a couple of generations the puffballs would have a total volume several times that of the Earth.

Greatest number of spores

NAME	**giant puffball** *Calvatia gigantea*
LOCATION	temperate regions
ABILITY	producing up to 20 trillion spores from just one fruiting body

Longest continuous incubation

NAME   **emperor penguin** *Aptenodytes forsteri*

LOCATION   Antarctica

ABILITY   incubating its egg through the worst of the Antarctic winter

This is the largest of the 17 species of penguins, and the only one that incubates through the Antarctic winter. Other birds, such as kiwis (see p.278) have longer incubation periods, but they share duties and leave their nests to feed. However, the male emperor penguin sits on an egg for the entire time that it takes to hatch: 62-67 days.

It all starts in late March or early April on the Antarctic sea ice where, after a summer feeding at sea, tens of thousands of the penguins gather to court and mate. By the time the eggs are laid, around 50 days have passed—and the penguins have had nothing to eat. Then, as the females all start walking back to the sea—which, because of the steadily growing sea ice, is farther away than it used to be—each male holds his egg on top of his feet and covers it with his brood pouch, a feathery flap of skin. The males stand huddled together through the depths of the Antarctic winter. It's dark or semidark, gales blow, blizzards rage, and temperatures may drop to -40˚ F (-40˚ C), not counting the windchill factor.

Two months pass, the eggs hatch, the chicks stay under the brood pouch, and in a few days the females start arriving. It's been around 120 days since the males have eaten anything other than snow, but they're finally free to go fishing. There's just one problem—the sea now can be as far as 62 miles (100 km) away.

Most artistic suitor

This little bird creates true works of art, each one a unique display. Instead of wearing colorful plumage, the male Vogelkop bowerbird courts females by appealing to their sense of aesthetics—a type of sexual selection through art. Even more extraordinary, different populations have developed different cultural styles.

In the Arfak, Tamrau, and Wandamen mountains a male starts with a "maypole"—a forest sapling—around which he weaves a conical stick hut up to four feet (1.2m) high and six feet (1.8m) wide, with an arch for an entrance. Outside he lays down a huge carpet of green moss on which he arranges his treasures, according to color and size—thousands of them, from berries and flowers to butterfly and beetle wings. Such an elaborate construction using fresh material needs constant renovation and guarding from other males, which can lead to almost year-round drudgery, but at least he has the safety of drab female colors instead of extravagant plumage. The more impressive the size and color of his bower, the more females will allow him to mate with them.

In the Kumawa and Fakfak mountains, however, females have developed a different artistic taste, perhaps determined by the local availability of objects for decoration. Here the males build dowdy towers decorated with dull objects such as snail shells and stones. It may be that, in years to come, selection of males for the artistic merits of their bowers—effectively, symbolic sexual symbols—will lead to one species, divided by the mountains, becoming two.

NAME **Vogelkop bowerbird**, or **brown gardener**, *Amblyornis inornatus*

LOCATION West Papua, Indonesian New Guinea

ABILITY building the most elaborately decorated structure created by a nonhuman animal

Most reproductive organs

NAME   **tapeworm** *Polygonoporus giganticus*
LOCATION   the gut
SIZE   depends on the size of the gut

Tapeworms are truly remarkable beasts that have been with us, intimately, since we were protohumans. Since they mostly inhabit the guts of animals, it means that the biggest guts are likely to have the biggest worms. Imagine, then, the excitement of a parasitologist who is presented with the challenge of dissecting the gut of a huge whale. Each type of tapeworm comprises a grappling headlike region (but a head with no mouth, eyes, or brain), designed primarily to attach to the gut of its host, followed by a type of a neck, which is the production line for the segments that make up the body.

Each segment is a self-contained reproduction kit, with testes and ovaries, that is able to absorb nutrients from juices in its host's gut. In the case of probably the longest tapeworm in the world, *Polygonoporus giganticus*, which lives in sperm whales, the body is reputed to be up to 131 ft. (40m) long with up to 45,000 segments. Since each segment has 4–14 sets of genitalia, that means a possible 630,000 sex organs. Its eggs are shed continually, millions a day. Luckily for their hosts, bigger tapeworms seem to be confined to one per gut. And since some tapeworms self-fertilize, there may be no problem about getting to meet a mate. It's lucky for us, though, that the adult phase of *Diphyllobothrium latum*, the largest tapeworm infecting humans (about ten million of us), rarely gets to be longer than 33 ft. (10m). Or is it really so lucky to have a 33 ft. long tapeworm inside of you?

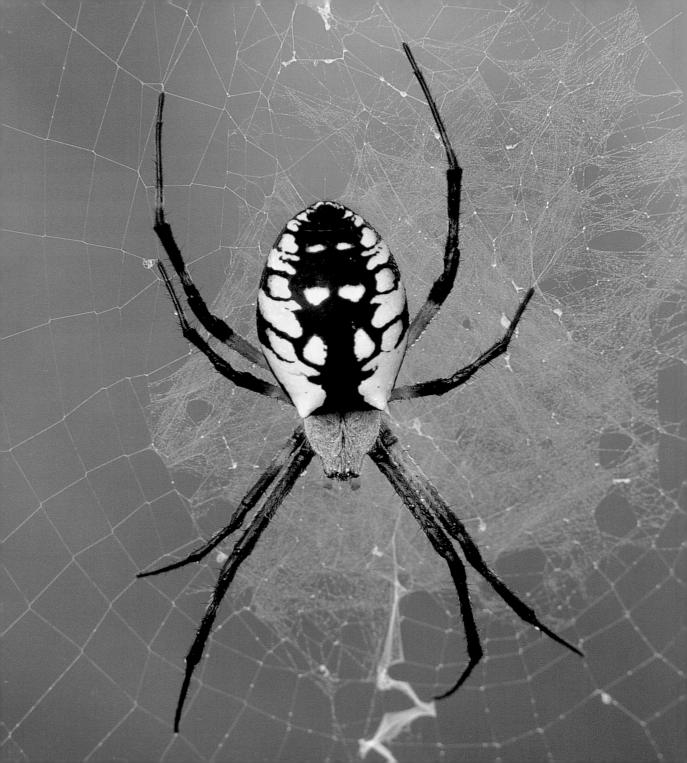

Most extreme mating ritual

Mating is a problem for most male spiders, since females are usually larger and extremely predatory. A male must therefore let his prospective mate know that her visitor comes bearing sperm, quite literally: he collects sperm from his genitals in each of his two pedipalps (the leglike appendages on each side of his mouth). These sexual "arms" have bulblike ends that fit, like a lock and key, into the genitalia of the female of his species.

Some males simply do a dance, make an appropriate signal, or offer a food parcel to earn enough time to mate. But those with short life spans, and therefore the prospect of mating only once, may have to sacrifice their lives to make sure that only their sperm fertilizes the female. When a male Australian red-back spider inserts his first pedipalp into the female, he somersaults over her so that his abdomen is directly over her mouth parts, presenting her with the opportunity to eat him. In an average 65 percent of matings, the male really is eaten, often when the female is underweight. But this, the ultimate sacrifice, allows him long enough to insert all of his sperm.

The male black-and-yellow garden spider goes one step further: he dies immediately after inserting his second pedipalp, which swells up inside the female. Neither males lining up to mate nor the female herself can dislodge his pedipalp, at least not for one quarter of an hour or so, by which time his sperm has done its job. And, of course, the most productive way for a female to dislodge the body hanging from her is to eat it.

NAME **black-and-yellow garden spider**
Argiope aurantia
LOCATION North and Central America
EVENT the male dies during mating, and the female then eats him

Strangest nuptial gift

If you were about to give a female one fifth of your body weight in sperm and nutrients, you'd want to make sure that she didn't die before her eggs were fertilized and deposited. And if her biggest danger was being eaten by a spider, you'd want to do something to prevent that from happening. This is how the scarlet-bodied wasp moth does it. He visits a plant, such as dog fennel and feeds on the poison (pyrrolizidine alkaloids) that it contains. The chemicals, which affect spiders and most other invertebrates, then make their way to the moth's pair of abdominal pouches, which are packed with fine, absorbent filaments. The male finds a female and, as part of his courtship, festoons her with these sticky, alkaloid-laden filaments. And just to make sure that she is protected, he also injects some of the alkaloid along with his sperm.

The dog fennel produces pyrrolizidine alkaloids to keep its leaves from being eaten by insect herbivores, but nature always seems to find a way around things—and the scarlet-bodied wasp moth has evolved an immunity to the poison. Spiders can detect the alkaloids when they examine a catch, and none of them wants to even touch a female scarlet-bodied wasp moth, much less eat her. Biologists proved this by putting a filament-festooned female on a spiderweb: the spider cut away the strands of its web that were holding her and released her, alive. This is the only known instance of a male insect making his mate invulnerable in such a way.

NAME **scarlet-bodied wasp moth**
Cosmosoma myrodora
LOCATION American tropics and subtropics
ABILITY the male makes the female invulnerable to spider attacks

In the 1830s, when scientists first became aware of this strange suborder of fish and started trawling specimens up from the depths—between 980 ft. (300m) and 13,000 ft. (4,000m)—there seemed to be some species that were entirely represented by females. Then someone examined a tumorlike growth that a few of these females seemed to have and was shocked to discover the male. He had fused onto her body, losing the ability to do anything independently. Even her circulatory system had extended into his body.

Since then a lot has been learned about some of the most extreme instances of sexual dimorphism (physical difference between the sexes—see p.310) in these fish. The male is generally less than one tenth of the female's size. In one of the largest species, *Ceratias holboelli*, for example, he measures 2.8 in. (7.3cm), and she's a massive 30. 3 in. (77cm).

The male starts life close to the surface as a tiny larva, metamorphoses into an adult, and then descends to a world that's dark and vast. He has to find a female of his species when there are probably only a few of them in a cubic mile. Fortunately, he's equipped with big nostrils, a big mouth, a pair of pincerlike teeth, one of the best senses of smell, and proportionately big eyes to look for the bioluminescent lure that a female uses to attract prey. If he's lucky, he finds a female, bites, and holds on. Then his eyes, nostrils—everything but his testicles—degenerate. And he's left with one simple role in life—to fertilize eggs.

Longest lasting love bite

NAME **deep-sea anglerfish** Ceratioidei family
LOCATION ocean depths
ABILITY sexual parasitism

These so-called double coconuts were impressing people long before the islands that they came from were ever discovered. They used to wash up on beaches around the Indian Ocean, and sailors would pick them up at sea. Before the discovery of the Seychelles in 1743, the general belief was that they were the fruit of a gigantic tree growing on the oceanbed—hence the name coco-de-mer or "nut-of-the-sea." Then, for a while, they were thought to come from the Maldives, which accounts for the scientific species name. Another theory was that they were from the Tree of Knowledge in the Garden of Eden, and the reason for that theory is pretty obvious: the striking resemblance to a woman from the waist to mid-thigh. Naturally, they were also thought to be aphrodisiacs.

The reality is pretty fantastic, too. The coco-de-mer palm tree grows incredibly slowly. The first leaf starts to appear nine months after germination begins, and the first flower can take another 60 years (there are male and female palms). The two-lobed fruit may take up to ten years to ripen, and it may be one hundred years before the tree reaches its full growth of about 100 ft. (30m). The tree's leaves, when they've finally grown, can be 20 ft. (6m) long. As for the nuts, they are edible and are a lot like coconuts. It's not likely, though, that anyone would be eating them nowadays. The tree is close to extinction, and the nuts, which are sometimes sold to botanical gardens, can cost as much as $1,500 (£800) each.

Biggest seed

NAME	**coco-de-mer**, or **Seychelles nut**, *Lodoicea maldivica*
LOCATION	Praslin and Curicusc islands, Seychelles
SIZE	the fruit can reach up to 19 in. (48cm) across and weigh more than 48 lbs. (22kg)

Greatest mass egg laying

NAME **olive ridley turtle** *Lepidochelys oliveacea*

LOCATION sandy beaches on the coast of Orissa, northeast India

EVENT 500,000 or more turtles may arrive to lay their eggs

At any time from December to March a huge flotilla of female olive ridley turtles arrives in the Bay of Bengal for an orgy of egg laying. This mass arrival has happened almost every year for thousands—possibly millions—of years, and the river delta beaches used comprise the biggest turtle rookery in the world. The largest recorded arrival there, in 1991, involved 610,000 olive ridley turtles.

The turtles travel possibly thousands of miles from their feeding grounds, mate offshore where the males congregate, and then haul themselves up the beaches at night to nest. They may even return a few weeks later to lay more eggs. Each turtle buries an average of 100–150 soft-shelled eggs in the sand. If the eggs are not eaten by predators, such as dogs

and crows, or washed away by beach erosion, they will hatch in around 45 days, and the babies will dig themselves out at night and run to the sea. But only one in one thou.s.nd will make it back to its birth beach as an adult.

In the past 14 years more than 120,000 olive ridley turtles are estimated to have drowned in the nets of trawlers fishing illegally within 6.2–12.4 miles (10–20km) of the coast during the nesting season. The area is also being developed, and there is a plan for a huge port complex close to one of the key breeding beaches. Conservationists worldwide are now pressing the state and national government to make sure that this—one of the greatest of the world's wildlife spectacles—gets the protection that it deserves.

The biggest egg actually belongs to the ostrich, weighing in at 2.2–39 lbs. (1–17.8kg). But it's only around 1 percent of the bird's weight. The largest eggs relative to body size and weight belong to storm petrels (also called Mother Carey's chicken), hummingbirds, and kiwis. The brown kiwi produces an egg that is so large in relation to its hen-sized body cavity that an x-ray of its body appears to be almost entirely egg. Although the egg is two thirds yoke and weighs up to 15–17 percent of its mother, it's laid after only 34 days— one of the fastest gestations of any bird.

Unsurprisingly, a kiwi usually only lays two to three eggs per year. Incubation, though, is one of the longest of any bird, taking up to 84 days, and is shared by the male (kiwis pair for life). When the baby kiwi breaks out of its egg, it is fully feathered and has a kick-start ration of yoke in its stomach. This almost mammal-like birth isn't surprising for a bird that has the lifestyle of a mammal. The kiwi is flightless with furlike feathers, nocturnal, lives in a burrow, has unbirdlike senses of hearing and smell (for locating food), catlike whiskers, and a body-temperature of 100°F (38°C), which is 36°F less than that of most birds. But the true origin of the enormous baby could be the kiwi's ancestry. It's a descendant of a lineage of giant birds that also gave rise to emus, and although its body is dwarf-sized by comparison, it has kept a virtually emu-sized egg.

Largest egg

NAME **brown kiwi** *Apteryx australis*
LOCATION New Zealand
SIZE up to 1 lb. (450g), filling up to 25 percent of its mother's body

Shortest mammal gestation

NAME **stripe-faced dunnart** *Sminthopsis macroura*
LOCATION Australia
TIME 9.5–11 days

Dunnarts are marsupial mammals. Like all marsupials, they produce tiny, undeveloped babies—embryos, almost—after a very short gestation period, or pregnancy. In fact, these newborns are so undeveloped that at first they breathe through their skin. Many marsupials have extremely short gestation periods, and only 12–13 days is common in a number of species, including the opossum and the long-nosed bandicoot. But the stripe-faced dunnart has the shortest gestation of all: as the breeding season progresses, it can shorten to just 9.5 days.

Like other mammals, marsupial embryos initially get all of their nutrition while they are in the womb from their mother's blood (via a placenta—an exchange organ). But while other mammalian babies remain in the womb for much longer, marsupials continue their development in their mother's external, abdominal pouch. And instead of receiving nutrition from their mother's blood, they fatten up on her milk. Unsurprisingly, marsupials often produce milk for much longer than other mammals. The stripe-faced dunnart continues to supply milk for more than two months—enough to feed as many as eight babies (she can produce more, but since she only has eight nipples, the excess babies will die). She carries her brood in her pouch as she scampers around after insects—until they get too big and must hitch a ride piggyback style if danger threatens.

Although not the most colorful of lizards, whiptails are the most genetically challenging. It took more than 100 years for biologists to realize that, in the Chihuahuan spotted whiptail and other species, no males existed. The all-female species were reproducing by "parthenogenesis"—without fertilization. It's now known that, among the higher animals, some other lizards, a few salamanders, the Brahminy blind snake, and a number of fish can also clone themselves. And individual snakes and domestic chickens and turkeys occasionally produce fertile eggs without sex, although the offspring are male and don't have all of their mother's genes.

Of the other species of whiptails that are parthenogenic, around half are triploid like the Chihuahuan whiptail, containing three sets of chromosomes. (Most animals are diploid, having only two sets: one coming from the mother, and the other coming from the father.) Biologists guess that the triploid state resulted from two whiptail species crossing, and their hybrid offspring then crossing with another whiptail. Strangely, one leftover from their past sexual lives remains: they can't give birth unless they pretend to mate— one female taking the role of a male. Biologists think that doing away with sex allowed them to reproduce themselves faster than their sexual relatives could and so spread far and wide rapidly—a good short-term strategy. In the long term, though, if conditions change dramatically, the theory says that these little clones may not have the genetic diversity to enable them to adapt and survive. But this could just prove to be sexual bias.

Sexiest immaculate conception

NAME	**Chihuahuan spotted whiptail**
	Cnemidophorus exsanguis
LOCATION	southwestern U.S. and Mexico
ABILITY	virgin birth

Most
fertile
animal

NAME   **aphids** Aphididae family
HABIT   plant suckers
ABILITY   a female is capable of producing
one billion clones in one year

For most of her life a female aphid produces identical copies of herself without sex. On average, she can do so every ten days, and since these clones are themselves pregnant before they are born—with their embryos containing embryos—a billionfold increase in an aphid population is theoretically possible in only one year.

Aphids have been around since plants first appeared on Earth. They are bugs with needlelike mouthparts for sucking up plant sap. If the sap is rich in nitrogen, for example, from artificial fertilizers, they grow even faster. Some ants milk aphids for their honeydew excretions and in return, the ants guard them and may even build shelters for them, store their eggs in the winter, or move them to new plants. These aphid populations may reproduce even more successfully, presumably because they are relatively free from insect predators such as ladybug, lacewing, and hoverfly larvae.

If the population gets too great for the host plant or the plant begins to die, they may start to give birth to winged progeny, and thus they can then spread themselves to new plants, making use of the wind. These winged individuals may produce sexual offspring on a new host species, the females of which lay eggs. But the size of a population is relative to the food supply, which in temperate areas declines as the winter approaches. And in a healthy environment the predator numbers are in tune with the cycles of their prey. So a rain of aphids is merely the stuff of nightmares.

The largest *tree* nest ever recorded was built in Florida by bald eagles. It had probably been added to over the years and was 9.4 ft. (2.9m) wide and 20 ft. (6m) deep. More normal bald eagles nests are much smaller—little more than stick piles compared to nests made by the true nest-building record holder.

However, a group of birds that are found in Australasia, called megapodes, use specially prepared incubation mounds to incubate their eggs. The average mound size of the chickenlike orange-footed megapode is around 11.5 ft. (3.5m) wide and 39 ft. (12m) high. The male and female, which usually pair for life, build it out of soil and organic materials.

The effort of constructing a mound, using hundreds of tons of small rocks, wood, soil, branches, and leaf litter, is immense and requires large and powerful feet—megapode means "big foot." A mound may be used over the years by many generations, growing larger and larger in the process—the record-beating 164 ft. (50m) wide specimen could have been in use for hundreds of years. The Sun heats the mound, but more importantly, decaying vegetation generates heat, and the temperature inside will be around 77–95°F (25–35°C). While the female lays her eggs—over a period of months— the male continues to tend to the mound, adding or removing nest materials to regulate the heat. It's similar to a giant, earthy compost pile, and, as the eggs hatch, the young must dig themselves out of the mound without any help.

Largest nest

NAME	**orange-footed megapode**, or **scrubfowl**, *Megapodius reinwardt*
LOCATION	southeast Asia, New Guinea, and northern Australia
SIZE	up to 164 ft. (50m) wide and 14.8 ft. (4.5m) high

Strangest
society

Naked mole rats organize themselves into the largest underground colonies (at least 75–80 individuals and up to 300) of any mammal. Conditions are hard, literally, as well as hot and humid, and so they have done away with fur and normal mammalian temperature control. The younger workers form the digging teams at the earth face searching for food, using their huge front teeth to gnaw through the hard soil until they find roots and tubers; they get rid of the soil by reversing along the tunnel and kicking it back up to the surface. A hierarchy exists, and when a subordinate meets a superior in a tunnel, she or he crouches so that the other can squeeze over the top. And like ants, there are guards—older, larger individuals.

Ruling all is the queen—one of the older animals (mole rats can live for 25 years or more), who fought for her position when the previous queen died. Like an ant queen, her abdomen has become elongated, enough to house up to 28 babies. She runs a small harem of males and constantly bullies the rest of the colony so that they are all too stressed to reproduce. There's sense in such organization. Where naked mole rats live, rain is unpredictable, and so there is no set season for reproduction. But underground, where there are always tubers to eat as long as there is a team to dig for them, it makes sense to have a social system where a colony behaves like one big organism, allowing a queen to breed all year round.

Clown fish are probably best known for their immunity to the stinging tentacles of sea anemones and the fact that particular fish live their entire lives among the tentacles of particular sea anemones. But it gets even more specific than that. Only a maximum of six clown fish will occupy one sea anemone, and only two of those will breed. The other four just reside there, tolerated by the breeding pair as long as they keep their places in the very strict clown fish hierarchy.

The dominant fish, and the largest, is the breeding female. Number two is the smaller breeding male, three is the next biggest, and so on. The fish control their own growth, and each fish in the pecking order is no more than 80 percent of the size of the one just above it. Any fish that gets ahead of itself by growing larger than its allotted size is evicted from the home anemone out into open water and, without any tentacles for protection (all the other appropriate anemones are usually occupied), to almost certain death.

So clown fish are very careful about their growth, and since they live for a long time, the little societies can survive without disturbance for decades. But when a fish does die, every fish below it is promoted and grows a little, and a new juvenile is allowed in. What happens, though, when the breeding female dies? Who takes her place? The breeding male, of course. Anything that can control its size can, it seems, control its sex, and so he turns himself into a she.

Most upwardly-mobile gender change

NAME	**clown anemonefish** *Amphiprion percula*
LOCATION	southern and western Pacific reefs
EVENT	males get promoted and become females

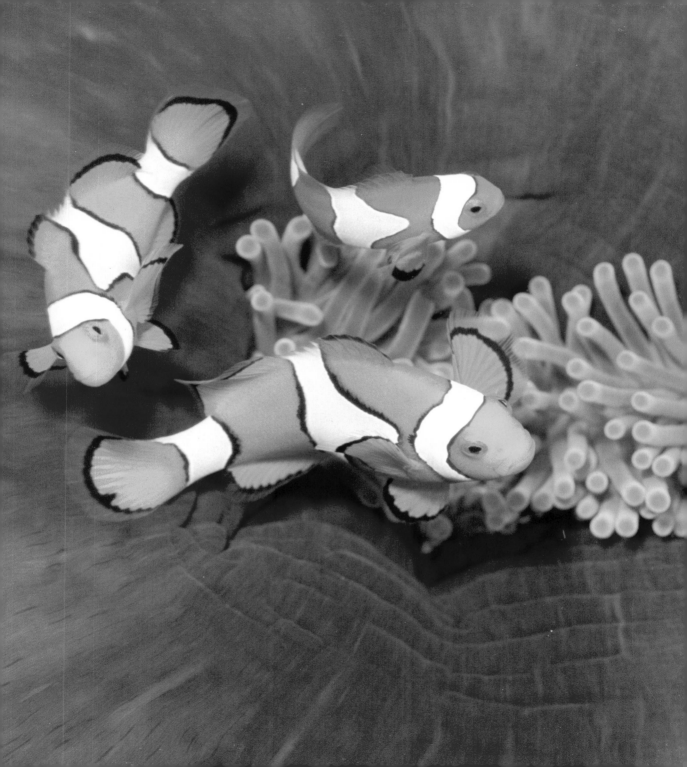

Most extreme reproductive act

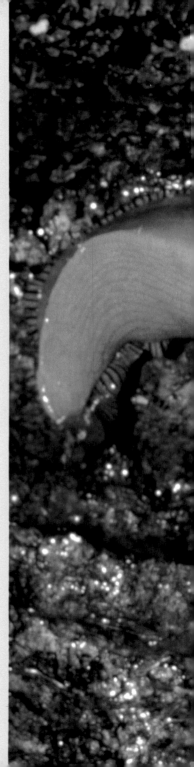

NAME **slender banana slug** *Ariolimax dolichophallus*
LOCATION California, U.S.
EVENT hermaphrodite partners' surprise ending

This is probably the most famous of all slugs. It is not only beautiful, for a slug, but it is also the official mascot of the University of California, Santa Cruz. The species is granted such respect in part for its exclusivity—it is only found in the redwood forests of central coastal California—but mainly for its intriguing sexual behavior. Rumors exist that the slender banana slug has the world's longest penis relative to body size, but that record is held by a goose barnacle. The penis, nevertheless, is sometimes as long as the slug's entire body (up to 7 in. (18cm), and the name *dolichophallus* does, in fact, mean "long penis."

The banana slug, like other slugs and snails, has both a female opening and a penis. After a long and sensual courtship two slugs copulate on a bed of slime, each one as both a male and a female simultaneously, for many hours at a time. But sometimes the female reproductive tract grips the penis so tightly that it can't be withdrawn. The only thing that the one entrapped slug can do is the act of apophally: gnaw off its penis (often helped by its partner). The advantage is that this prevents the castrated slug from wasting resources on providing other slugs with sperm, and so, the theory goes, it will divert more energy into growing its fertilized eggs. And, of course, its partner will have had a nutritious, post-coital snack.

Biggest group-spawning event

Corals have devised a spectacular way to maximize their chances of cross-fertilizing: the mass release of sperm and eggs. The most dramatic of these procreation events occurs on the Great Barrier Reef—the largest coral reef ecosystem in the world. Mass spawning of reefs in the area occurs over a few nights after a full moon, in October, November, or sometimes December, and involves more than 140 different coral species. This spectacular phenomenon is synchronized by three main environmental cues: warming of the seawater in the spring, which triggers the eggs and sperm to mature (most corals are hermaphrodites), the lunar cycle (peak spawning is four to six nights after a full moon), and darkness.

The mass spawning creates a pink and white underwater snowstorm, as billions of coral polyps release their small egg-and-sperm bundles. These float to the surface, where the sperm fertilize the eggs. The fish have a bonanza, but there are only so many eggs they can eat, and some eggs contain deterrent chemicals. Within hours of fertilization, an egg develops into an embryo, which becomes a small larva and floats away searching for a free place on the seabed to settle and start a coral colony through multiplication of itself. Mass coral spawning is one of the world's great wonders, repeated by some but not all coral reefs around the world. The slick formed by the eggs on the surface can be seen from space; surprisingly, though, the event was only recorded for the first time in 1981.

NAME **reef corals**—more than 140 species

LOCATION Great Barrier Reef, Australia

EVENT sperm and eggs are released by all the corals at the same time on the same night

Most colorful male

NAME	**sapphire copepods** *Sapphirina* species
LOCATION	subtropical and tropical oceans
ABILITY	glittering with the most beautiful iridescence

Sapphire copepods are a tiny fraction of an inch, wood louse-shaped crustaceans (small relations of shrimps and crabs), and form part of the sea's rich plankton. Color plays a critically important role in their lives. While the free-swimming female stays as inconspicuous as she can, the rather more sedentary, iridescent-blue male literally sparkles—reflecting a rainbow of colors whenever the Sun strikes him. His body covering, or cuticle, contains layer upon layer of crystal platelets that act like prisms, and the iridescence comes from the diffraction and refraction of white light as it shines through them.

Intriguingly, different males—even though they belong to the same *Sapphirina* species—have prismatic crystals of different thickness that produce different colors. Why, though, no one

knows for sure. The males are not free-swimming but live inside of transparent salps (jet-propelled, barrel-shaped animals), eating away at their insides, and are dependent on the salps for transportation. The likelihood is that the parasitic males have to flash to signal their whereabouts to the females. In fact, their potential mates have much better developed eyes than they do and can see in 3D, presumably to help them locate flashing males.

Once a female finds a suitably colorful flasher and indicates that she likes him, he swims out of his salp, and they mate. But the downside for the male is that his host may have since moved on, and so he is left homeless and with no food supply until he can find another likely looking salp to sneak up to and into.

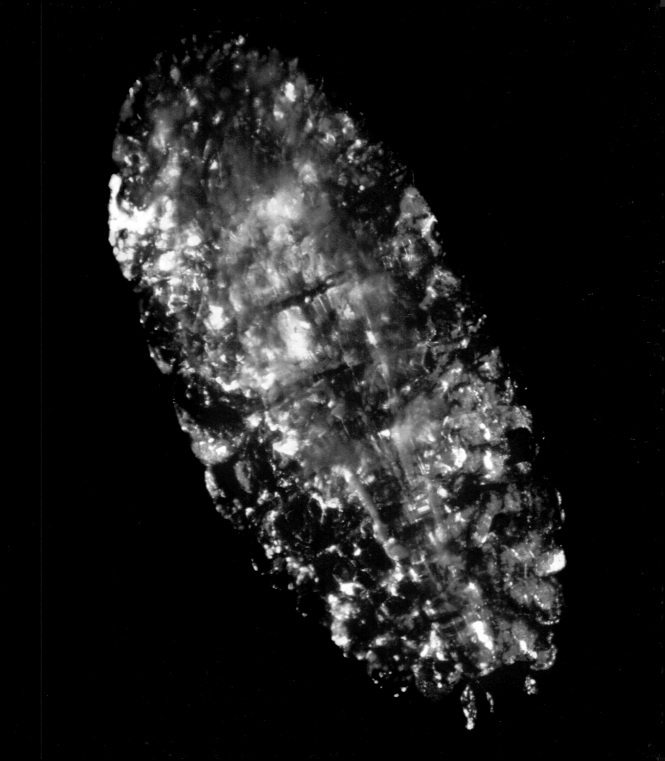

Sexiest animal

For bonobos mating is not simply a method of reproduction but an essential part of social intercourse used to avoid conflict, make peace, and create a harmonious society. Sexual contact between females is common and maintains close friendships. It also enables young females migrating to a new group (males stay in the group they were born into) to bond with established females. In fact, bonobo society—unlike chimpanzee society—is female centered, and female friends will join forces to keep aggressive males under control. Males don't form such close bonds. The closest relationship for a male is with his mother, whom he remains attached to throughout her life. His status depends on hers.

Bonobos occasionally walk upright (although they spend much of their lives in the trees) and display great intelligence—in captivity they can even communicate using basic human language.

Females usually become pregnant only every six to seven years, and youngsters are nursed until they are at least five years old. But such slow reproduction is disastrous for bonobos. Logging of their rainforest home and rampant poaching has fragmented their population and may have reduced their numbers to fewer than 10,000 individuals.

NAME **bonobos**, or **pygmy chimpanzees**
Pan paniscus
LOCATION Democratic Republic of the Congo, south of the Congo River
ABILITY uses creative mating techniques to build relationships and make peace

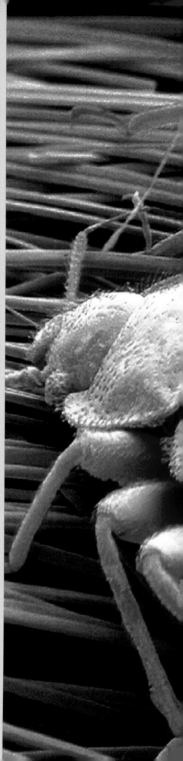

Most traumatic insemination

NAME	**common bedbug** *Cimex lectularius*
LOCATION	possibly in a bed near you
EVENT	the male stabs the female to inject his sperm

A male bedbug not only has mouthparts for sucking up human blood but also a sperm-delivery needle (an evolved penis) for stabbing female bedbugs. Sperm is injected directly into the female's blood and swims up to her ovaries. Unfortunately, though, the stab wounds may let in infections. This rather violent technique has evolved as a way of bypassing the female's genital passage and therefore any control she has over the timing of conception, as well as any antisperm devices. (Female animals are believed to have such devices, since if a male isn't staying around to help raise the offspring, a female may prefer to widen the genetic options by mating with several males, even selecting whose sperm to use to father her offspring.)

"Traumatic extragenital insemination" seems like an extreme example of males getting on top in the sexual-conflict arms race. But females have evolved a counteraction. Judging by the scars on a female, a copulating male's body position often results in his organ piercing into the region of her fifth segment. If he stabs to the right, his organ slides into a notch that directs it into a type of holding pocket, complete with spermicidal cells that can kill both unwanted sperm and microbes. Females mated in this way live longer and produce more eggs. A male, though, is on the watch for promiscuous females. He has taste sensors on his stabbing organ, and if these detect sperm, he injects much fewer of his own, saving his resources for his real desire: virgins.

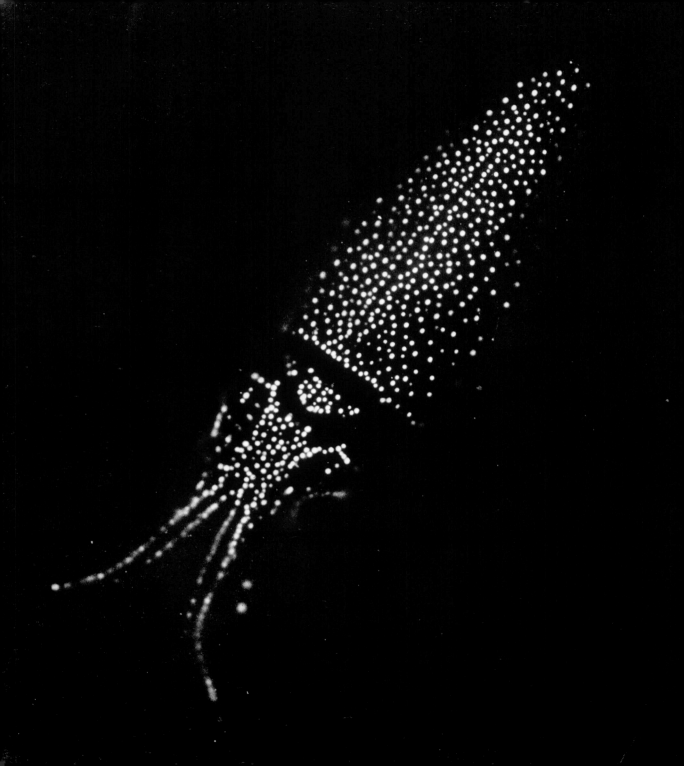

Sparkliest animal

At night from March to May Toyama Bay on the west coast of the Japanese island of Honshu lights up. The water flashes, sparkles, shimmers, and glows, and the spectacle attracts visitors from all over Japan and from around the world. Toyama prefecture has even declared the bay to be a Special Natural Monument. What happens there is the annual spawning of the firefly squid. There are hundreds of thousands of small squid, each one 1.6–2.4 in. (4–6cm) long, capable of producing light from hundreds of special light-producing patches of tissue on their skin called photophores. Animals that produce light are called bioluminescent, and Japanese firefly squid are probably the most exuberantly bioluminescent creatures on the planet.

When they're spawning (producing eggs and sperm), the squid are undoubtedly using their lighting as a form of sexual communication, but when they're not spawning, they have other uses for it.

They spend most of the year out in the open sea, either in the Pacific or in the Sea of Japan. During the day they stay at a depth of 650–1,970 ft. (200–600m), and at night they rise to the surface. It's thought that they use the photophores to break up their outlines, confuse predators, and possibly even lure in prey. Cephalopods (squid, octopus, and cuttlefish) are very visual animals, but the firefly squid, which probably can see in color, may be the most visual of all.

NAME	**Japanese firefly squid** *Watasenia scintillans*
LOCATION	western Pacific
ABILITY	putting on a mass underwater display of bioluminescence

Most imaginative use of dung

NAME   **burrowing owl** *Athene cunicularia*

LOCATION   grasslands of North, Central, and South America

ABILITY   using dung as bait to catch beetles

A burrowing owl uses a burrow as a place to shelter, rest, hide, and nest in. Its scientific name *cunicularia* means "little miner," but it likes a ready-made burrow best. When spring comes, a male burrowing owl will select or dig a burrow—ideally in a short-grass habitat with insects and small rodents that can easily be seen and caught at ground level. It may then collect dung—especially shredded cow or horse dung (in times past, it might have been buffalo or antelope dung)—and place it inside of the nest chamber and around the entrance.

It has been suggested that this might mask the smell of the eggs and the young to protect them from predators such as badgers. But a more nutritious use has been discovered. The regurgitated pellets of owls with dung in their burrows contain ten times more dung beetles than those of owls that don't use dung. So the dung beetles, whose own nest activity involves finding dung to roll into balls on which to lay their eggs, are almost certainly being lured in by the male owls.

The dung provides "meals-on-wheels" for the incubating females and, indeed, for the males themselves, who spend much of their time guarding the nest burrows and so have relatively little opportunity to hunt. To date, the only other records of animal use of dung for nonfood purposes are as fuel by humans and as nest lining by a bird or two, leaving owls as the record holders for ingenuity with dung.

Oldest juvenile

NAME **olm**, **blind cave salamander**, or **human fish**, *Proteus anguinus*
LOCATION underground streams in limestone caves in mountains along the Adriatic Sea, from Trieste in Italy to Montenegro
ABILITY existing in a juvenile form for more than 50 years

A "neotenic" lifestyle—remaining an eternal tadpole, complete with gills and fins—suits a group of salamander-type amphibians that have chosen an almost completely aquatic life. In the case of the Mexican axolotl, found in the wild only in two lakes, transformation into a proper land-roaming salamander can occur if the water dries up. But the rest of the neotenic amphibians (including North American mud puppies and sirens) stay as juveniles for life.

One of the strangest is the eel-like olm, which is blind and has tiny limbs and translucent, baby-pink skin (although a black population exists in Slovenia). It lives in dark, underground caves and hunts insect larvae and other small creatures by smell and vibrations. Being blind isn't a problem, as it smells its way around and can probably navigate using Earth's magnetic field. When times are hard, it can go without food for years, probably in a state of torpor.

Although it keeps its larval fin and gills throughout life (rumored to be up to 100 years), when it gets to be seven or so, it becomes sexually mature. The female normally lays eggs, but under certain, perhaps dry, conditions, she is believed to give birth to live young. Biologists think that the olm took to a subterranean life after the last ice age, retreating to the water in limestone karstic formations. Since it never surfaced again, it didn't bother to lose its tadpole fins and gills and so remains an eternal juvenile.

Most quarrelsome siblings

NAME   **spotted hyena cubs** *Crocuta crocuta*
LOCATION   African savanna
ABILITY   siblings start fighting as soon as they are born

It's tough in the natural world, and even siblings cannot always rely on one another. The first-hatched pups of sand tiger sharks, for instance, quickly use up the food in their yolk sacs and start dining on their smaller brothers and sisters until there are just two dominant pups left, one in each uterus. Spotted hyena cubs don't actually eat each other, but they start fighting—often quite viciously—the moment they are born.

There are usually two cubs in a litter, and they both have very sharp teeth. One often dies of starvation or battle-induced injuries, especially if they are the same sex, but if they are different sexes, they usually both survive. They're not actually fighting over milk (although if the weaker one survives, it accepts that it is subordinate and allows its litter mate to exclude it from nursing) but are preparing for adult life.

Their success in hunting big animals is largely a result of clan cooperation, but when it comes to feeding on the kill, spotted hyenas compete vigorously. How much an individual can eat is determined by where it is in the dominance hierarchy. Females are dominant over all the males, and then there is a hierarchy within each sex. Critically, in the case of cubs whose mothers are low-ranking, the amount of food they are allowed to eat determines to a great extent on whether they will survive beyond adolescence. Sibling rivalry is therefore a matter of life or death in more ways than one.

Males and females inevitably have different mating needs and strategies, and that can give them different shapes, colors, or sizes. This is known as sexual dimorphism, and it's common throughout the animal kingdom. In the case of most dimorphic animals, the two sexes are similar enough to be recognizable at least as the same species. However, it would be hard to know, just by looking, that a male and female blanket octopus had the slightest connection with each other. The female can be 6.6 ft. (2m) long and weigh 22 lbs. (10kg). The male grows to all of 0.9 in. (2.4cm) and almost 8.8 oz. (300g). And the reason, essentially, is that they inhabit the open sea, a boundless world that is controlled by currents.

For the female this means being as large as possible so that she can produce as many eggs as possible, increasing the chances that at least a few will survive without being eaten. For the male growing too big would be pointless. All he needs to do is manufacture sperm and focus on searching the ocean for a female. If he finds one—a very big if—he inserts an arm into her gill cavity, and the arm fills with sperm and breaks off. He leaves it there and, having completed his life's mission, floats off and dies. In fact, scientists had only ever seen dead male blanket octopuses until 2003, when the first living one ever seen was caught in the beam of a flashlight.

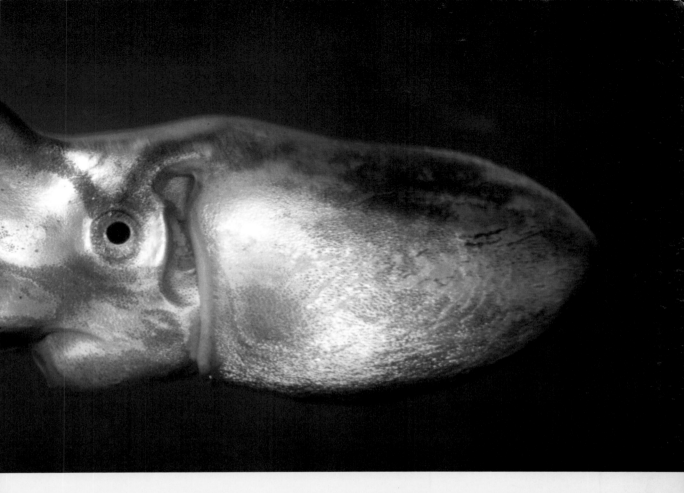

Greatest sex divide

NAME	**blanket octopus** *Tremoctopus violaceus*
LOCATION	warm seas worldwide
SIZE	the female is 100 times larger than the male

No mother carries eggs or babies for longer than she needs to, but the alpine salamander does it for up to three years and two months—the longest gestation that has ever been recorded for any vertebrate (animal with a backbone). The salamander's average pregnancy is two years—considerably longer than the 20-22-month gestation of an Asian elephant—and yet, surprisingly, the female is no more than 5.6 in. (14cm) long. So why does she do it? The clue is in the alpine location: damp but cold, with only a short period in the spring and summer when she can be active. In fact, the higher she lives, the longer her pregnancy is. The likelihood is that, as the altitude increases, it gets more and more difficult for her to find enough food to sustain herself *and* grow babies in just one season.

Strangely, unlike most other amphibians, she doesn't lay eggs. Instead, she nurtures her young internally, in her two oviducts (only mammals have a uterus). She produces 20 or more eggs in each oviduct, but only one egg develops in each one—the others provide food for the two growing larvae. Even stranger, the youngsters have special teeth to scrape their mother's reproductive tract, giving them enough nourishment to last through their long gestation and to metamorphose from larvae into fully-formed salamanders before birth. Once they are born, though, their mother's work is done, and they are left to fend for themselves.

Longest pregnancy

NAME	**alpine salamander** *Salamandra atra*
LOCATION	high mountain forest of alpine regions of Europe
TIME	up to 38 months

A nest functions to keep the eggs together and to cushion, insulate, shelter, or conceal them, and nest building is as varied as a bird's ingenuity and the sites and material available. Even the most conservative nest may contain something unusual if nature (or humanity) provides. Common waxbills put carnivore droppings in their nests, possibly as defense against predators. Some wrens and flycatchers use the cast skins of snakes and lizards, and in their absence, they'll collect plastic or cellophane. For other species, if mud isn't available for pottery work, cow dung might do. And if there's a shortage of animal hair for weaving, wool, string, or even nylon tights and metal shavings are perfectly acceptable.

The birds that consistently choose the strangest nesting material, however, are the corvids: crows, magpies, jays, and the like. Most are omnivorous, meaning they will have a go at eating most things, which makes them perfectly adapted for urban life with its endless supply of items to satisfy both hunger and their insatiable curiosity. They particularly love shiny things, and all types of eye-catching objects have been found in their nests, from rings and watches to aluminum foil. In Tokyo, though, where crows have really taken to urban life, some have replaced their normal nesting material—large sticks—with shiny coat hangers, which they scavenge from garbage bags. They have even perfected a way of carrying them in flight. There's a big problem for humans though: when the crows choose to nest on electricity pylons and drop a hanger or two, there is one massive electrical short-circuit.

Strangest nesting material

NAME	**common crow**, (hashiboso-garasu)
	Corvus corone
LOCATION	Tokyo, Japan
ABILITY	making use of urban garbage, in particular coat hangers, as nest material

Acknowledgments

With grateful thanks to the following scientists and experts who gave their valuable time to advise and comment.

Dr. Duur Aanen, University of Copenhagen, Denmark; **Hilary Aikman,** Department of Conservation, Wellington, New Zealand; **Professor Kellar Autumn,** Lewis & Clark College, Oregon, U.S.; **Dr. Jason Baker,** National Marine Fisheries Service, Honolulu, U.S.; **Bat Conservation International,** Texas, U.S.; **Professor Ernest C. Bernard,** University of Tennessee, Knoxville, U.S.; **Kat Bolstad,** Auckland University of Technology, New Zealand; **Dr. Elizabeth Brainerd,** University of Massachusetts, Amherst, U.S.; **Dr. Mark Brazil,** Rakuno Gakuen University, Hokkaido, Japan; **Professor Lincoln Brower,** Sweet Briar College, Virginia, U.S.; **Professor Malcolm Burrows,** Department of Zoology, University of Cambridge, U.K.; **Dr. Eliot Bush,** University of Chicago, Illinois, U.S.; **Dr. Peter Buston,** National Center for Ecological Analysis and Synthesis, University of Santa Barbara, California, U.S.; **Dr. John A. Byers,** University of Idaho, Moscow, U.S. (author of *Built for Speed*); **Dr. John A. Byers,** Western Cotton Research Laboratory, United States Department of Agriculture—Agricultural Research Service, Arizona, U.S.; **Dr. Tom Cade,** The Peregrine Fund, Boise, Idaho, U.S.; **Dr. Janine Caira,** University of Connecticut, Storrs, U.S.; **Dr. S. Craig Cary,** Center for Marine Genomics, University of Delaware, U.S.; **Rogan Colbourne,** Department of Conservation, Wellington, New Zealand; **Professor William E. Conner,** Wake Forest University, Winston-Salem, North Carolina, U.S.; **Dr. Iain Couzin,** University of Oxford, U.K.; **Dr. David Croft,** Arid Zone Field Station, University of New South Wales, Fowlers Gap, Australia; **Dr. Alistair J. Cullum,** Creighton University, Omaha, Nebraska, U.S.; **Dr. John W. Daly,** Scientist Emeritus, National Institute of Health, Bethesda, Maryland, U.S.; **Professor Frans B. M. de Waal,** Living Links Center, Emory University, Atlanta, Georgia, U.S. (author of *Bonobo: The Forgotten Ape*); **Dr. W Richard J Dean,** Percy FitzPatrick Institute of African Ornithology, University of Cape Town, South Africa; **Dr. Mark W. Denny,** Hopkins Marine Station, Stanford University, California, U.S.; **Dr. Andr.ew Derocher,** Edmonton, University of Alberta, Canada; **Professor Christopher R. Dickman,** Institute of Wildlife Research and School of Biological Sciences, University of Sydney, New South Wales, Australia; **Dr. Steve Diver,** University of Arkansas, Fayetteville, U.S.; **Professor Johan T. du Toit,** Director, Mammal Research Institute, University of Pretoria, South Africa; **Professor Tom Eisner,** Section of Neurobiology and Behavior, Cornell University, U.S.; **Dr. Mark Erdmann,** Senior advisor, Conservation International's Indonesia Marine Program; **Professor Frank E. Fish,** West Chester University, U.S.; **Professor Charles Fisher,** The Pennsylvania State University, U.S.; **Dr. Diana Fisher,** School of Botany and Zoology, Australian National University, Canberra, Australia; **Dr. Matthias Foellmer,** St Mary's University, Halifax, Nova Scotia, Canada; **Dr. Scott C. France,** University of Louisiana, Lafayette, U.S.; **Dr. Laurence Frank,** Museum of Vertebrate Zoology, University of California, Berkeley, U.S.; **Dr. Douglas S. Fudge,** University of Guelph, Ontario, Canada; **Dr. Lisa Ganser,** University of Miami, Florida, U.S.; **Dr. Randy Gaugler,** Rutgers University, New Jersey, U.S.; **Professor Ray Gibson,** John Moores University, Liverpool, U.K.; **Dr. Cole Gilbert,** Department of Entomology, Cornell University, U.S.; **Dr. Matthew R. Gilligan,** Savannah State University, Georgia, U.S.; **Dr. Ross L. Goldingay,** Southern Cross University, Lismore, New South Wales, Australia; **Professor Gordon C. Grigg,** University of Queensland, Brisbane, Australia; **Dr. Paul D. Haemig,** Umeå University, Sweden; **Professor Bill Hansson,** Department of Crop Science, Swedish University of Agricultural Sciences, Alnarp, Sweden; **Alice B. Harper,** Aptos, California, U.S.; **Professor Peter Harrison,** Southern Cross University, New South Wales, Australia; **Dr. S. Blair Hedges,** Pennsylvania State University, U.S.; **Dr. Wilbert Hetterscheid,** Curator, Botanical Gardens, Wageningen University, The Netherlands; **Dr. A. Rus Hoelzel,** University of Durham, U.K.; **Dr. Carl D. Hopkins,** Cornell University, Ithaca, New York, U.S.; **Dr. David J. Horne,** Queen Mary, University of London, U.K.; **Dr. Ludwig Huber,** University of Vienna, Austria; **Dr. Michael Huffman,** Kyoto University, Primate Research Unit, Japan; **Professor George Hughes,** University of Bristol, U.K.; **Dr. Marina Hurley,** *Writing Clear Science*, Richmond South, Victoria, Australia; **Claudia Ihl,** University of Alaska, Institute of Arctic Biology, Fairbanks, Alaska, U.S.; **Paul Jansen,** Department of Conservation, Wellington, New Zealand; **Professor Jennifer U. M. Jarvis,** Zoology Department, University of Cape Town, South Africa; **Dr. Judy Jernstedt,** University of California, Davis, U.S.; **Dr. Reinhard Jetter,** University of British Columbia, Vancouver, Canada; **Professor**

Dale W Johnson, University of Nevada, Reno, U.S.; **Dr. Darryl N. Jones,** University of Queensland, Brisbane, Australia; **Dr. Peter J. Jones,** School of Biology, University of Edinburgh, U.K.; **Dr. Stephen M. Kajiura,** FAU Elasmobranch Research Laboratory Biological Sciences, Florida Atlantic University, U.S.; **Professor Ilan Karplus,** Aquaculture Research Unit, Ministry of Agriculture, Bet Dagan, Israel; **Professor Masashi Kawasaki,** University of Virginia, Charlottesville, U.S.; **Dr. Rob Kay,** Laboratory of Molecular Biology, Medical Research Council, Cambridge, UK; **Professor Laurent Keller,** Department of Ecology and Evolution, University of Lau.s.nne, Switzerland; **Professor Keith Kendrick,** Gresham Professor of Physic, Laboratory of Cognitive and Behavioral Neuroscience, The Babraham Institute, Cambridge, U.K.; **Dr. Ian Kitching,** The Natural History Museum, London, U.K.; **Cor Kwant,** Amsterdam, The Netherlands; **Dr. Jack R. Layne Jr.,** Slippery Rock University, Pennsylvania, U.S.; **Jonathan Leeming** www.scorpions.co.za; **Ashley Leiman,** The Orangutan Foundation, London; **Professor Douglas Levey,** University of Florida, Gainesville, U.S.; **A. Jasmyn J. Lynch,** University of Queensland, Gatton, Australia; **Dr. Andrew L. Mac,** Wildlife Conservation Society, Goroka, Papua New Guinea; **Dr. Todd J. McWhorter,** College of Agricultural & Life Sciences, University of Wisconsin, Madison, U.S.; **Dr. Christopher D. Marshall,** Department of Marine Biology, Texas A&M University, Galveston, Texas, U.S.; **Professor Justin Marshall,** Sensory Ecology Laboratory, Vision Touch and Hearing Research Centre, University of Queensland, Brisbane, Australia; **Chris Mattison,** Sheffield, U.K.; **Dr. Mette Mauritzen,** Institute of Marine Research, Bergen, Norway; **Don Merton,** Department of Conservation, Wellington, New Zealand; **Dr. Jay Meyers,** Northern Arizona University, Flagstaff, U.S.; **Dr. Geoff Monteith,** Queensland Museum, Brisbane, Australia; **Dr. John Morrissey,** Hofstra University Marine Laboratory, New York, U.S.; **Dr. Ulrich G. Mueller,** University of Texas, Austin, U.S.; **Dr. Phil Myers,** Museum of Zoology, University of Michigan, Ann Arbor, U.S.; **Dr. Mark Norman,** Museum Victoria, Melbourne, Australia; **Tim Osborne,** Kori Bustard Farming, Outjo, Namibia; **Peter Parks,** Image Quest Marine, Witney, Oxfordshire, U.K.; **Dr. Julian C. Partridge,** University of Bristol, U.K.; **Professor Jakob Parzefall,** Zoologica Institute and Zoological Museum, University of Hamburg, Germany; **Dr. Irene Pepperberg,** Brandeis University, Massachusetts, U.S.; **Professor Eric R. Pianka,** The University of Texas, Austin, U.S.; **Professor Theodore W. Pietsch,** University of Washington, Seattle, U.S.; **Robin Prytherch,** Bristol, U.K.; **Gordon Ramel,** the Earthlife Web; **Dr. Simon M. Reader,** Utrecht University, The Netherlands; **Ian Redmond,** Bristol, U.K.; **Dr. Eileen C. Rees,** Wildfowl & Wetlands Trust, Martin Mere, Lancashire, U.K.;

Dr. Barry A. Rice, University of California, Davis, U.S./International Carnivorous Plant Society; **Dr. Klaus Riede,** Zoological Research Institute and Museum Alexander Koenig (ZFMK), Bonn, Germany; **Dr. Stephen Rossiter,** Queen Mary College, University of London, U.K.; **Royal Botanic Garden Edinburgh,** U.K.; **Dr. Roy T. Sawyer,** Biopharm Ltd., Wales, U.K.; **Dr. John S. Scheibe,** Southeast Missouri State University, Cape Girardeau, U.S.; **Dr. Scott L. Schliebe,** Marine Mammals Management, U.S. Fish and Wildlife Service, Anchorage, Alaska, U.S.; **Jason C. Seitz,** Florida Program for Shark Research, Florida Museum of Natural History, University of Florida, Gainesville, U.S.; **Professor Lynne Selwood,** University of Melbourne, Australia; **Dr. Jamie Seymour,** James Cook University, Cairns, Queensland, Australia; **Professor Craig Sharp,** Brunel University, Middlesex, U.K.; **Colin Shawyer,** Wheathampstead, Hertfordshire, U.K.; **Professor Rick Shine,** University of Sydney, New South Wales, Australia; **Dr. Stephen C. Sillett,** Humboldt State University, Arcata, California, U.S.; **Dr. Colin Simpfendorfer,** Center for Shark Research, Mote Marine Laboratory, Florida, U.S.; **Dr. Jack T. Sites Jr.,** Brigham Young University, Provo, Utah, U.S.; **Nigel Sitwell,** Galapagos Conservation Trust, London, U.K.; **Dr. Mike Siva-Jothy,** University of Sheffield, U.K.; **Dr. David G. Smith,** National Museum of Natural History, Smithsonian Institution, Washington, D.C., U.S.; **Dr. Peter Speare,** Australian Institute of Marine Science, Townsville, Queensland, Australia; **Dr. Brian Spooner,** Royal Botanic Gardens Kew, Richmond, Surrey, U.K.; **Dr. Adam P. Summers,** University of California, Irvine, U.S.; **Professor Sidney L. Tamm,** Biology Department, Boston University, Boston, Massachusetts, U.S.; **Dr. Andrew Taylor,** Wildlife Unit, Veterinary Sciences, University of Pretoria, South Africa; **Dr. Robert J. Thomas,** University of Cardiff, U.K.; **Dr. Richard Thorington,** National Museum of Natural History, Smithsonian Institution, Washington, D.C., U.S.; **Dr. Michael Tyler,** University of Adelaide, South Australia; **Dr. J Albert C. Uy,** Syracuse University, U.S.; **Dr. Richard Vari,** National Museum of Natural History, Smithsonian Institution, Washington, D.C., U.S.; **Dr. Peter Vukusic,** Exeter University, Devon, U.K.; **H. J. Walker Jr.,** Scripps Institution of Oceanography, University of California, San Diego, U.S.; **Dr. Simon J. Ward,** University of Melbourne, Parkville, Victoria, Australia; **William Watson,** NOAA Southwest Fisheries Science Center, La Jolla, California, U.S.; **Dr. Grahame Webb,** Wildlife Management International, Sanderson, Northern Territory, Australia/Chairman IUCN-SSC Crocodile Specialist Group; **Dr. Alexander Weir,** Behavioral Ecology Research Group, University of Oxford, U.K.; **Dr. Barbara Wienecke,** Australian Antarctic Division, Tasmania, Australia; **Professor Oystein Wiig,** University of Oslo, Norway; **Dr. Alex Wilson,** Center for Insect Science, University of Arizona, Tucson, Arizona, U.S.; **Professor Peter Wirtz,** Portugal; **Dr. Jeffrey J. Wood,** Royal Botanic Gardens, Kew, U.K.; **Dr. William F. Wood,** Humboldt State University, California, U.S.; **Dr. Jennifer L.Wortham,** University of Tampa, Florida, U.S; **Dr. Don E. Wilson,** Smithsonian Institution, Washington, D.C., U.S.

And a very special thank you to **Helen Brocklehurst** and **Rachel Ashton** for their invaluable advice, support, enthusiasm, and encouragement from beginning to end.

Index